Discovering Astronomy
Activities Manual and Kit

Third Edition

by

D1537293

R. Robert Robbins
University of Texas, Austin

William H. Jefferys
University of Texas, Austin

Stephen J. Shawl
University of Kansas

John Wiley & Sons, Inc.
New York   Chichester   Brisbane   Toronto   Singapore

Preface

This Activity Kit contains a variety of activities that, while integrated into *Discovering Astronomy*, may also be used independently with minimal or no additional instruction. We therefore assume that the reader has an astronomy book for reference that includes pictures of the Sun, Moon, planets, nebulae, and galaxies. The Activity Kit includes three general kinds of activities: observations of the sky; experiments using lenses, a diffraction gratings, and polaroid filters; and activities using photographs to make measurements and form conclusions. An appendix contains more extensive telescopic observing activities. All these activities complement the Discovery Activities in the text, which are generally shorter and do not require any of the materials in the kit.

The activities allow the reader to understand better the measurement process and the processes scientists use in forming conclusions. For example, readers learn the importance of repeated measurements, the meaning of averages, and the errors associated with the measurement process. In this way, readers learn that science is not exact and that *all* measurements have an associated uncertainty. They must then form conclusions based on imperfect and incomplete data, just as scientists do. Thus, while on a simple level, actively involved readers are learning first hand what science is. If readers completing these activities have an improved understanding of the scientific process, we will have met our most important goal.

The observing activities include observations using a cross-staff, quadrant, naked-eye, and telescope. The simple equipment will provide data similar in quality to those obtained during the renaissance period during which great advances in astronomy were made using instruments of similar design.

The second type of activity includes examining the properties of lenses, diffraction, polarization, and spectra. These activities provide opportunities for understanding a variety of phenomena within our every day environment.

The final group of activities uses various materials included in the kit itself, such as photographs of the Virgo and Hercules clusters of galaxies, photographs of Jupiter that are used to determine the planet's rotation period, photographs and a spectrum of the Crab Nebula that allow determination of its distance, and so on. Astronomical research is often done from archival photographs, and these activities further illustrate, using hands-on methods, how scientific results are obtained.

Many of the activities include answer sheets that not only guide the reader in the collection of data, but provide a standard form to make grading easier and help instructors provide students with feedback.

Not all the activities provided would be done during a typical semester. Those chosen depend on the goals and interests of the individual instructor. Instructors may want to refer to the Instructor's Manual that accompanies *Discovering Astronomy* for further ideas on the use of the activities in a course setting.

We are pleased to acknowledge the editorial help of Pui Szeto.

Table of Contents

Kit Activity 3-1

Construction and Use
of an Astronomical Cross-staff

When you have completed this activity, you should be able to do the following:

• Make a cross-staff and use it to measure angles.

• Average measurements and determine the amount of error in a measurement.

• Explain the purpose of finding an average and the meaning of the standard deviation (which is approximated by a simplified calculation called the Snedecor rough check) as a measure of uncertainty.

• Measure the angular size of an object from several distances, and plot a graph of the results with error bars.

• Use a graph to determine the value of one of the variables given the value of the other.

• Compare an observed relationship with a theoretical one, and judge whether the two are in good agreement with one another.

Use the answer sheet at the end of the Activity.

The cross-staff is a simple instrument for the measurement of angles and angular sizes. In using a cross-staff to map the sky, you will obtain accuracy roughly equivalent to that obtained by most of the world's astronomers up until the naked-eye work of Tycho Brahe (1546 - 1601) and the subsequent work of Galileo in 1609, when he first observed the sky with a simple telescope.

In this activity, you will practice making measurements. In the next chapter, you will make measurements of the sky.

Part I
Discovering the Science of Astronomy

CONSTRUCTING THE CROSS-STAFF
AND PRACTICING WITH IT

To assemble your cross-staff, take the pattern page containing the CROSSPIECE FOR CROSS-STAFF from your activity kit. Push out both the large pattern piece and also the small GRATING HOLDER, discarding all the rest of the leftover cardboard.

Staple the GRATING HOLDER to its indicated location on the big crosspiece before you do any folding. Then fold the scored sides to make right angles. Flap A will attach to the place that says "paper clip or staple to A" using either large paper clips or a *small* stapler.

Punch out the two pieces marked PUNCH OUT and slide the device onto the meter stick, making sure that the side of the crosspiece with the grating holder on it faces toward the zero end of the meter stick. Your assembled cross-staff should resemble **Kit Figure 3-1-1**. (NOTE: If a meter stick is unavailable, you can use a yard stick instead.)

Now practice using your cross-staff to measure some angles. Make two marks 1 meter (or yard) apart on a piece of paper hung on a wall or blackboard. Stand 4 meters (or yards) away from the marks. Rest the zero end of the meter stick against your cheekbone and sight along the length of the stick, sliding the crosspiece back and forth until the left-hand mark on the wall is lined up with the left-hand edge of the wide sight on the cross-staff, and the right-hand mark is lined up with the right-hand edge of the wide sight (**Kit Figure 3-1-2**). (The wide sights are 4-inches wide; the medium and small sights are 2 and 1 inches wide, respectively.)

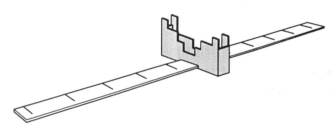

Kit Figure 3-1-1 The assembled cross-staff.

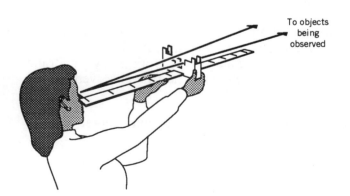

To objects being observed

Kit Figure 3-1-2 Using the cross-staff to measure angles. (Not to scale)

Part I
Discovering the Science of Astronomy

After lining everything up as well as you can, remove the stick from your cheek and read off the value on the meter stick where the vertical crosspiece on the cross-staff closest to your eye intersects the meter stick. Read the measurement to the nearest 1 mm. You should get a reading of approximately 42 cm on the meter stick to within 1 or 2 cm. (With a yard stick, measure 1 yard from a distance of 4 yards; you should get a reading around 16.5 in.)

It is now necessary to convert the 42 cm (16.5 in) reading into an angle. One method of doing this is to use the nomogram of **Kit Figure 3-1-3**. (A nomogram is a graphical device that can be used instead of a formula to relate several quantities to each other.) To use it, place it on a flat surface and lay a straightedge from your cross-staff reading (left-hand scale) through the sight you used (center scale); where the straightedge intersects the angle scale on the right, you will find the angle you have measured. If all goes well, you should be able to determine that the 42 cm reading corresponds, with the wide sight, to an angle of about 14°. Thus, the angle between two marks 1 meter apart, viewed from a distance of 4 meters, is about 14°.

The angle can also be computed using a formula known as the angular size formula:

$$\text{Angular Size} = 57.3° \times \frac{\text{true size}}{\text{distance}}.$$

For the example above, we can take the "linear size" to be the width of the wide sight (10 cm), and the "distance" is the reading on the meter stick (42 cm). We then have

$$\text{Angular Size} = 57.3° \times \frac{10 \text{ cm}}{42 \text{ cm}} = 13.6 \text{ degrees}.$$

One other method to find the angle from the measurement is to use the graphs in **Kit Figure 3-1-4**. You can use these graphs for any angle shown; they are mathematically rigorous. Furthermore, they clearly show how the uncertainty in the angle increases dramatically as the curve becomes steeper with increasing angle.

Kit Figure 3-1-5 shows a drawing of the measurement you just made and illustrates how the cross-staff works on a principle of similar triangles. The closer the crosspiece is to your eye, the larger the angle you can see. If the crosspiece had only one set of notches, the angles could be marked directly on the stick. However, three pairs of sights allow the measurement of a wider range of angles.

Again, stand 4 meters or yards from the wall and remeasure the angle between the marks, this time using the medium-width sights. As before, convert the reading on the meter stick to an angle, but this time be sure that you use the middle mark on the nomogram. You will get a different reading on the meter stick, but the angle you read from the nomogram should again be about 14°. This makes sense, since you are measuring the same angle as before.

In general, you should choose the sight that is most convenient and easiest for you to use. Usually you will use the wide sight to measure large angles and the narrow sight to measure small angles.

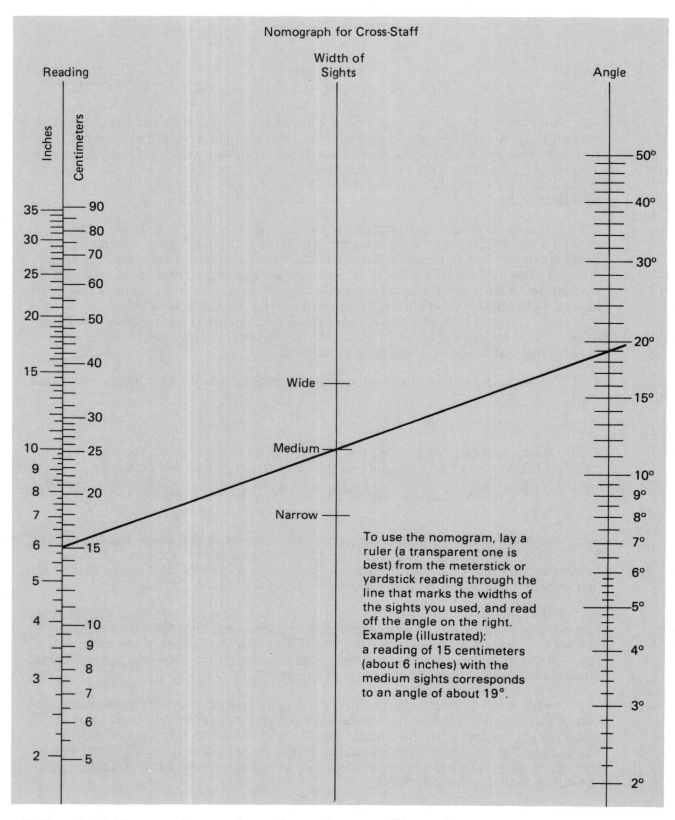

Kit Figure 3-1-3 Nomogram for converting readings on the cross-staff into angles.

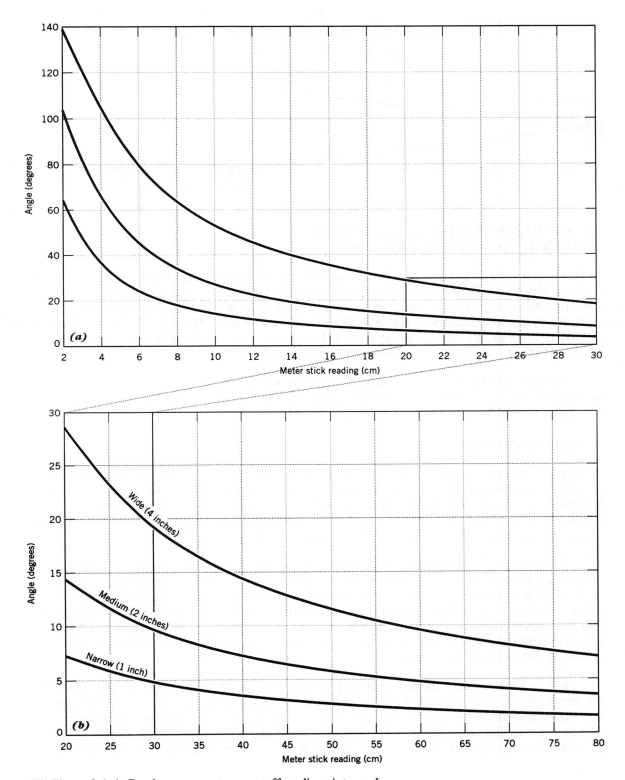

Kit Figure 3-1-4 Graphs to convert cross-staff readings into angle.

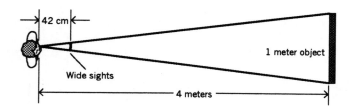

42 cm

Wide sights

1 meter object

4 meters

Kit Figure 3-1-5 The similar triangles
principle of the cross-staff.

MEASUREMENT OF ANGULAR SIZES
AND ERRORS OF MEASUREMENT

In this experiment, you will use the cross-staff to measure the angular size of an object at various distances. To obtain the most accurate results, you will make repeated measurements and average them, because this technique tends to reduce the errors inherent in any measurement process. This fundamental activity, prerequisite to all other measurements in this book, can be done at home. If you are in a class, your instructor might choose to set it up in a convenient laboratory area.

At the end of this activity is a data sheet on which you should record your data and answer the activity questions. It is recommended that you photocopy this page and use the copy, so that you will have a backup in the event of mistakes.

A. Making Measurements

As before, set up two marks on the wall, 1 meter apart. Stand 2 meters from the marks and measure the angle between them, reading the meter stick to the nearest millimeter. Compute the angle with the formula, the nomogram, or the graphs; if you use the nomogram, try to read the angle to the nearest tenth of a degree. (If you have a yard stick, set up the marks 1 yard apart and stand 2 yards away.)

It is difficult to read to an accuracy of 0.1°. In fact, with each step you will have some uncertainty. Any measurement process inevitably involves uncertainties, even those made with the most sophisticated equipment. Part of the purpose of this experiment is to make you aware of methods for dealing with certain types of uncertainties. Never assume that measurements are "exact" simply because they were made with an elaborate and expensive apparatus. For this reason, scientists must constantly be concerned about the uncertainties in their measurements.

Record the measurement you have just made of the angle between the two marks in the first table of your answer sheet. Now move the sliding piece on the cross-staff; once you have moved the crosspiece, the next measurement you make will be completely independent of the first. Remeasure the same angle and record the result. Repeat this process three more times, for a total of five independent measurements of the same angle.

B. Random Errors

If you have made each measurement independently of the others, there will probably be some variation between one measurement and the next. This is to be expected, since it is impossible to do everything in exactly the same way each time. For example, when you slide the crosspiece back and forth, sometimes you set it too short and sometimes too long; when you stand 2 meters from the marks, it will never be exactly 2 meters, and so on. You should expect these random errors, and so you should not artificially try to force all your measurements to come out the same. Make each measurement as accurately as you can. DO NOT ROUND THEM OFF.

• **Kit Inquiry 3-1a** In addition to the two sources of errors given in the text, give at least three additional sources of *random* errors in your experiment. (Answer this and other questions to follow in the space provided on your answer sheet.)

C. Averaging Measurements

Random errors will tend to cancel each other out if you make a series of measurements and average them together. To determine the average, add the five measurements you have just made and divide the total by five (the number of measurements you made). Record your results. In general, the average value is a more accurate approximation of the true value of the quantity you are trying to measure than any of the individual measurements.

D. Estimating the Random Error of Your Measurements

We will estimate the random error through a simplified method called the Snedecor Rough Check. Examine your measurements and find the largest and smallest values of the angle. Determine the range of your measurements (by subtracting the smallest measurement from the largest) and then compute your approximate error by dividing your range by 2 (if you made 5 measurements). In general, you divide by $\sqrt{N-1}$, where N is the number of measurements. For example, if you had found that the range of the five measurements you made was equal to 2°, the uncertainty would be about 1°. If the value you determined for the angle was 28°, then the customary way to express the result of your measurement would be to say that the value of the angle was 28° ± 1.

• **Kit Inquiry 3-1b** What was the uncertainty of the five measurements that you just made? (Record your result.)

While the nomogram is a convenient device for converting your measurement to an angle, there is some degree of uncertainty in using it.

• **Kit Inquiry 3-1c** If you can read the angle to 0.1 of the distance between tick marks on the nomogram, how accurately can you read the angle scale for an angle of 5°? 20°? 50°?

A set of readings that averages to 15 cm might have an uncertainty of ±0.5 cm. From the nomogram, assuming use of the medium sight, we can determine that this ±0.5 cm uncertainty translates into a ±3/4° uncertainty in the angle.

- **Kit Inquiry 3-1d** What is the uncertainty in the angle for measurements made at 40 and 80 cm assuming ±0.5 cm in the reading?

- **Kit Inquiry 3-1e** What produces the larger source of uncertainty in your angle, the uncertainty of your reading, or the conversion to the angle using the nomogram? In other words, does the use of the nomogram itself add significantly to the uncertainty in the computed angle? How about when you use the graph?

E. Measurement from Greater Distances

Return to the two marks and stand 4 meters (or 4 yards) away from them. As before, make five independent measurements of the angle; be sure to record your answers. Calculate the average and uncertainty and record these also. Then repeat the entire observing procedure from a distance of 8 and 12 meters (or 8 and 12 yards). At this point, you will have made and recorded a total of 20 measurements.

Graph your measured results on the graph paper accompanying your answer sheet as follows: The horizontal axis represents the distance that you were standing from the marks, and the vertical axis represents the angle you measured. Plot the four average values you computed on the graph. For example, above number 2 on the horizontal scale, put a point at the appropriate height by finding the value on the vertical scale corresponding to the average angle measured from a distance of 2 meters.

In addition, draw a horizontal dash above and below each point you plotted. The distance above or below the point should be equal to the uncertainty you computed for that point. Connect the two dashes with a vertical line, as shown in **Kit Figure 3-1-6**. The vertical line between the two horizontal dashes is called an **error bar**. It gives a visual indication of the amount of confidence to be given to each point. The smaller the error bar, the more confidence you have in the accuracy of the point.

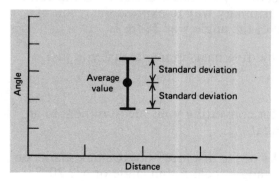

Kit Figure 3-1-6 Error bars give a visual indication of the confidence we place in each data point.

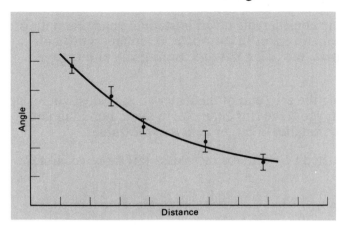

Kit Figure 3-1-7 The correct way to connect points.

When you have plotted all the required points on the graph, take a pencil and, freehand, draw a smooth curve that connects all the average values on your curve (**Kit Figure 3-1-7**). Do not use straight lines, since you have been measuring a quantity that varies smoothly, not in jerks. Now the resulting graph visually represents all the information you have obtained, describing the variation of the angle with distance as you determined it experimentally.

• **Kit Inquiry 3-1f** Now that you have the graph, you can use it to predict observational results for other distances. For example, use your graph to predict the angular size of a 1-meter object measured from a distance of 10 meters. (**Hint:** Locate the number 10 on the horizontal axis and draw a vertical line upward until it intersects the curve. Then draw a horizontal line from this point toward the left until it intersects the angular size axis.)

F. Systematic Errors in Cross-staff Measurements

In addition to the random errors just discussed, there is also a possibility of systematic errors. This type of error is far more difficult to detect and correct than random errors, and so it is a greater threat to accuracy. Systematic errors are effects that cause a series of measurements to be consistently too small or too large. For example, if the sights on the cross-staff were closer together than they are supposed to be, you would have to put the slide consistently closer to the eye end of the meter stick than otherwise necessary in order to measure a given angle. This would make the reading on the meter stick too small.

• **Kit Inquiry 3-1g** If the sights on the cross-staff were closer together than they should be, would the measured angle be smaller or larger than the true value? (**Hint:** Check the nomogram for converting meter stick readings to angle measurements.)

Unless you are clever enough to suspect that a systematic error is present, or unless you have an independent check on a series of measurements, it is possible for a systematic error to go undetected for a long period of time. There are many examples of this in astronomy. Scientists must expend considerable effort in studying and attempting to account for the various kinds of systematic errors in their data. It remains a difficult task, however, because there are no general procedures for measuring and eliminating systematic errors.

• **Kit Inquiry 3-1h** If, instead of pressing the zero end of the meter stick against your cheek, you were to hold it farther from your face (for example, to avoid bumping your glasses), would the measured angles be smaller or larger than the true value?

• **Kit Inquiry 3-1i** Would the changes listed below make the cross-staff more accurate? Why?

(a) Making the entire instrument larger (that is, larger sliding piece, longer stick, and so on).

(b) Having finer divisions marked on the meter stick.

(c) Marking a scale of angles directly on the meter stick, so that it is not necessary to read the graph.

G. Comparison of Theory and Observation

We introduced a formula near the beginning of the activity that made it possible to compute the angular size of an object, given its true size and distance. The test of any theory is its consistency with observations, assuming that the observations are accurate. Using the theoretical formula (Section 3.3), compute the angular size that you should have observed for each of the distances you used, namely 2, 4, 8, and 12 meters. Plot these points in a different color on your graph and connect them with a smooth curve.

• **Kit Inquiry 3-1j** Are the theoretical values within the error bars on the graph? If not, which one(s) aren't? Explain why this can happen.

Activity 3-1   Construction and Use of an Astronomical Cross-Staff

---

Answer Sheet   NAME _____

1. Record your measurements of the angle between the marks, from each of the four distances, in the tables below. The first table is to record the readings from meterstick or yardstick of your cross-staff, and the second is to record your conversions of these readings into angles.

DISTANCE (meters/yards)	2	4	8	12
1", 2", or 4" edges?	_____	_____	_____	_____
Measurement #1	_____	_____	_____	_____
Measurement #2	_____	_____	_____	_____
Measurement #3	_____	_____	_____	_____
Measurement #4	_____	_____	_____	_____
Measurement #5	_____	_____	_____	_____

Angles below should be the values from calculation or the nomogram, in decimal degrees.

	2	4	8	12
Angle #1	_____	_____	_____	_____
Angle #2	_____	_____	_____	_____
Angle #3	_____	_____	_____	_____
Angle #4	_____	_____	_____	_____
Angle #5	_____	_____	_____	_____
Average	_____	_____	_____	_____
Error	± _____	± _____	± _____	± _____
Theoretical value from formula of Section 3.3.	_____	_____	_____	_____

(continued on next page)

2. Plot your measurements of angle versus distance on the graph below.

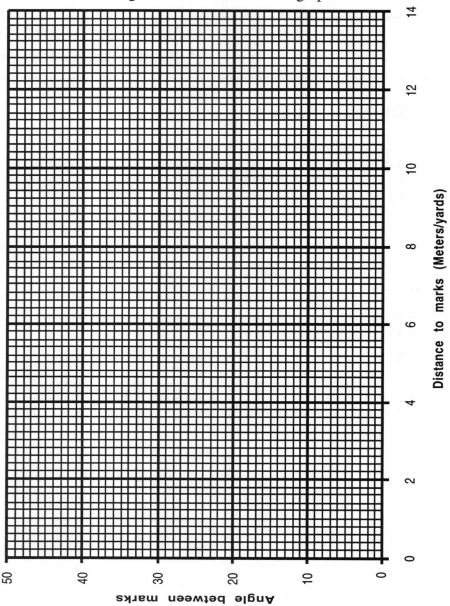

Distance to marks (Meters/yards)

Angle between marks

3. Answers to the Kit Inquiries on a separate piece of paper.

Kit Activity 3-2

Constructing and Using an Astronomical Quadrant

After completing this activity you should be able to do the following:

• Make a quadrant and use it to measure altitudes.

• Use the quadrant to measure the height of a building and determine the uncertainty of the measurement.

Use the answer sheet at the end of the Activity.

The quadrant is an instrument for measuring the altitudes of celestial objects. The altitude of an object is the angle between the horizon and the object, as shown in **Kit Figure 3-2-1**. It should be apparent that the altitude of the horizon itself is 0°, whereas that of the zenith, or point overhead, is 90°. *Do not confuse the term "altitude" with the same word that is used in aviation to mean the distance of a plane above the ground. They are entirely different concepts!*

The quadrant you will make is similar in function to the large mural quadrant used by the Danish astronomer Tycho Brahe. Brahe's exquisitely accurate naked-eye observations played a crucial role in overthrowing the 2000-year-old Ptolemaic theory of the universe and substituting the modern Sun-centered solar system. Even today, Tycho's observations could not be improved on without a telescope. The instrument you will construct will be similar to his in operation, although not as large, and it will be hand-held rather than mounted on a solid base. It will be a good example, however, of a type of instrument used to observe the heavens for over a thousand years.

In this activity, you will practice making measurements. In the next chapter, you will make measurements on the sky.

ASSEMBLING AND USING THE QUADRANT

Remove the quadrant pattern (marked QUADRANT) from your experiment kit and attach it to a straight stick with thumbtacks, as shown in **Kit Figure 3-2-2**. The stick should be at least 6 inches in length, and it should have some thickness to it to help attach the pattern and to aid your eye in sighting along it. In other words, it should be a board rather than a stick. (You can also use tape with a paperback book instead of a stick; it is very effective. Using the meter stick or yard stick does not work very well.)

The top of the pattern should be aligned carefully with the top edge of the stick, and

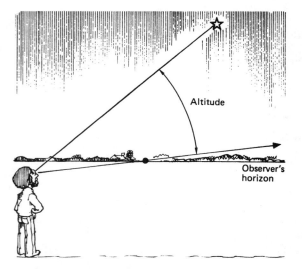

Kit Figure 3-2-1 The definition of altitude as an angle above the horizon.

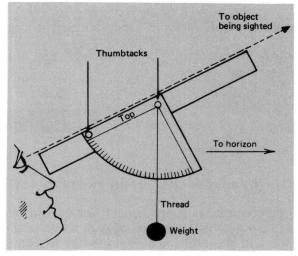

Kit Figure 3-2-2 The quadrant. The thread and weight define a vertical line, and the altitude is measured with respect to it.

thumbtacks should be pushed through the two crosses on the top of the quadrant. If the pattern is attached correctly, then the quadrant should give a reading of 0° when the distant horizon is sighted. Suspend a small weight from the front thumbtack on a foot-long piece of white thread or fishing line. Do not use ordinary string, which is too thick and will not swing freely. The weight should not be too light or it will blow in the wind and introduce errors. The weight of half a dozen keys is about right.

To use the quadrant, sight along the top edge of the stick at the object whose altitude is to be measured. Wait for the weighted thread to stop swinging. At this time, the Earth's gravity causes it to define a line that points exactly toward the zenith and the nadir, the point directly underneath the observer. Carefully, so as not to disturb the position of the thread, trap it against the scale with your finger or thumb, remove the quadrant from your eye, and read off the value where the thread crosses the protractor scale. Try to estimate your reading to the nearest tenth of a degree.

To practice using your quadrant, pick an object that is at a small altitude (less than 10°) and another at a large altitude (50° or greater). You could stand outside a building and observe a first-story window and compare it with a window much higher up. Do each of the following: measure the altitude of each object you have chosen for a total of five times using the quadrant, compute the average altitude of each, and estimate the error of each measurement in degrees. Now compute the percent error of each measurement:

$$\text{percent error} = \frac{\text{estimated error}}{\text{measured value}} \times 100 .$$

For example, if the altitude were 30° and its error were 1°, the percent error would be $1/30 \times 100 = 0.033 \times 100$ or about 3%.

• **Kit Inquiry 3-2a** Does the percent error increase or decrease with larger measured angles?

Measuring the Height of a Building with the Quadrant

Select a relatively tall building (more than two or three floors) that is surrounded by a sufficient amount of level ground so that you can stand off from it at a distance approximately equal to the height of the building. Back away from the building, taking occasional sightings with the quadrant on the top of the building, until the altitude of the building is as close to 45° as you can get it (see **Kit Figure 3-2-3**). Mark the spot where you are standing and carefully pace off the distance to the building. Multiplying the number of paces by the average length of your pace will give the distance to the building. To measure your pace, step off 10 paces and measure the distance with your meter stick; then divide this distance by 10. If you pace in a deliberate and uniform manner, you will be able to measure distances with fair accuracy this way. (Record the resulting distances.)

Now repeat this process one more time (stepping back from the building, marking the point where you stand, and pacing off the distance), and again record your result. If your answers are reasonably close to your first results, you need not repeat them again.

It is now possible to estimate the height of the building. Kit Figure 3-2-3 shows that, when the altitude of the building is just 45°, the distance AB from the building is equal to the distance AC, the amount of the building that is above eye level.

• **Kit Inquiry 3-2b** How should you figure in the effect of your height (refer to Kit Figure 3-2-3)?

• **Kit Inquiry 3-2c** Determine the height of the building.

If you are in a class, your instructor may want everyone to measure the same building, so that the measurements can be compared with each other and averaged together as an additional example of the method of averaging.

Like the cross-staff, the quadrant is imperfect and measurements made with it are subject to errors, both random and systematic.

- **Kit Inquiry 3-2d** What kind of error (random or systematic) would each of the following effects introduce?

     (a) The quadrant is poorly assembled, as in **Kit Figure 3-2-4.**

     (b) The weight swings back and forth.

     (c) You are unable to sight the object in exactly the same way each time.

     (d) The wind blows toward you and moves the weight slightly toward you.

     (e) Suggest three other possible sources of error.

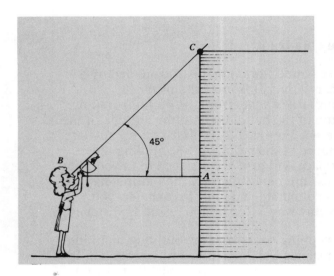

Kit Figure 3-2-3 Measuring the height of a building

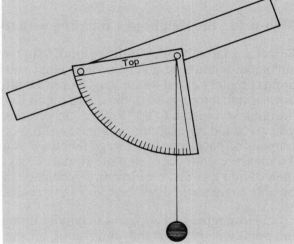

Kit Figure 3-2-4 A poorly assembled quadrant.

Activity 3-2 — Constructing and Using an Astronomical Quadrant

Answer Sheet   NAME _____

1. Record the five measurements you made of a small- and large-altitude angle below. Then calculate the average and estimate the error and percent error. Record these numbers in the space provided.

	Small Altitude	Large Altitude
Measurement #1	_____	_____
Measurement #2	_____	_____
Measurement #3	_____	_____
Measurement #4	_____	_____
Measurement #5	_____	_____
Average ± Error	_____ ± _____	_____ ± _____
Percent Error	_____	_____

Answer to Inquiry #3-2a

What building did you measure the height of? _____

Length of ten of your paces: _____ Length of one pace: _____

Number of paces to the building: _____

Inquiry 3-2b:

Inquiry 3-2c: Height of building (corrected for your height): _____

Inquiry 3-2d:

Part I
Discovering the Science of Astronomy

Kit Activity 4-1

Measuring Northern Stars
with Quadrant and Cross-staff

When you have completed this activity, you should be able to do the following:

• Use a sky map as a guide to find the principal constellations and stars in the northern part of the sky at any time of year.

• Determine your latitude from the altitude of the polestar.

• Use your quadrant and cross-staff to measure the positions and motions of stars in the sky, and interpret the measurements.

Use the answer sheet at the end of the Activity.

On a clear night, find a location as far from interfering lights as possible. A horizon that is relatively free of obscurations such as buildings and trees, especially in the north, is also desirable. Consult your star maps to see what stars you should expect to find in the north for the time of year it is, and find them in the sky. Make a sketch showing your horizon and its landmarks and the rough positions of the stars you can see.

At least twice during the evening use your instruments to make the observations indicated below. Be sure to make at least three (and preferably five) independent measurements of each angle and, if they appear consistent, average them together to minimize random errors. In addition, be sure to estimate the errors of the observations. For your data to be usable, you will need to let nearly two hours elapse between your two sets of measurements; *your results will be much better if you let more time pass* (thus giving the sky time to "move").

In addition to Polaris, you will need to choose two easily recognized stars to the east of Polaris and two west. For example, in September, you would probably want to choose two stars in the Big Dipper and two in Cassiopeia. Note that you do not need to know the names of the stars you use, but *you must learn the patterns well enough to guarantee that when you take your second set of measurements, you are measuring the same stars.* For each star chosen, you should measure:

Part I
Discovering the Science of Astronomy

1. The altitude of the star, with the quadrant.

2. The angular separation between Polaris and the star, with the cross-staff.

3. The *position angle* of the star with respect to Polaris, with the quadrant (see the following discussion).

The **position angle** of a star is the angle made by a line from Polaris to the star with respect to a line parallel to the horizon.[1] **Kit Figure 4-1-1** shows how to measure the position angle of a star with respect to Polaris by using the quadrant sideways and reading off the angle as indicated. For some stars, the quadrant string will not fall on the scale of the instrument. In these cases, you need to turn the quadrant around so that the scale is facing *away* from you and record the *negative* of the angle shown on the scale. What you will be finding from these data is how much the angle changes and in what direction.

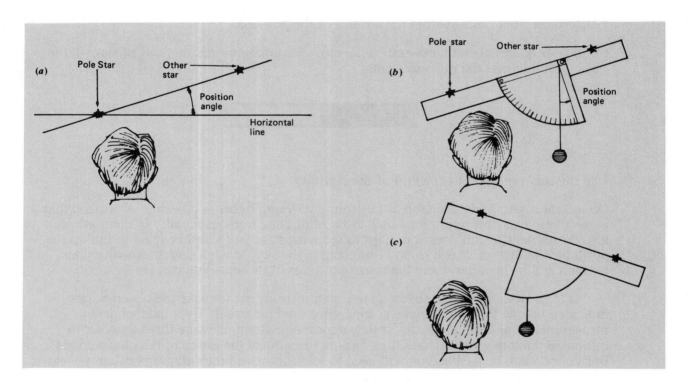

Kit Figure 4-1-1 Position angle. (*a*) Definition. (*b*) Measurement. (*c*) Measurement if the star on the left is higher (negative position angle).

_____

[1]While this is a different definition of position angle than is customary in astronomy, it is easier to apply with these instruments.

Part I
Discovering the Science of Astronomy

Record all angles and the times of observation on your answer sheet. Wait for nearly two hours (preferably more), and then repeat all your observations again. After making your observations, plot them on graph paper so that you can more easily visualize the motions of the stars in the northern sky. If you follow steps 1 and 2 in the next paragraph carefully, you should have no difficulty in plotting the measurements you have made.

1. Locate Polaris on your graph by plotting a point on the line above the north point of the horizon at the altitude that you measured for Polaris. **Kit Figure 4-1-2** shows this done for an altitude of 45°.

2. Using your measurements, locate each of the other stars you observed relative to Polaris. For example, suppose the measured altitude of a star was 50°, and its distance from Polaris was found to be 30°. **Kit Figure 4-1-2** shows how to plot this star on the graph. Using a compass, or a pencil tied to a piece of string, measure 30° on the axis of the graph. Strike an arc 30° in radius, using Polaris as the center. The star will be located at that point on the arc that has an altitude of 50°. NOTE: *You must use a compass or string to plot the point; you cannot just mark your point above the 30 mark on the horizontal scale.*

- **Kit Inquiry 4-1a** Why is it incorrect simply to mark this data point above 30°E on the horizontal scale?

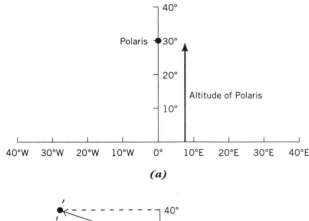

(a)

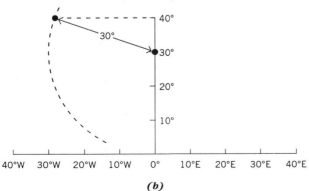

(b)

Kit Figure 4-1-2 Locating Polaris (*a*) and a second star (*b*) on the graph. In this example, the second star is located at an altitude of 40° (horizontal line) and a distance of 30° from the pole (arc of circle). The intersection of these two lines fixes the position of the star on the graph.

- 21 -

After completing your graph, answer the following questions by referring to it and to your recorded data.

• **Kit Inquiry 4-1b** *On the basis of your observations*, which stars (those to the east or those to the west) have altitudes that increase? Decrease? Maintain constant altitudes?

• **Kit Inquiry 4-1c** *On the basis of your observations*, describe in your own words the motions of the northern sky. Do the angular distances of stars you measured from Polaris increase, decrease, or remain the same?

• **Kit Inquiry 4-1d** On the basis of your observations of the position angles of the various stars, how many degrees does the position angle of a star change in an hour? According to your measurements, how long will it take the stars to rotate a full 360°? (You have just determined the Earth's sidereal period!)

• **Kit Inquiry 4-1e** On the basis of your observations in this activity, what was the latitude of the place where you made your measurements?

Between your first and second set of measurements, you can refer to your star maps and find other northern stars. If you are also doing Kit Activity 4-2, you can carry out those measurements while waiting for the necessary time interval to pass before making your second set of measurements for the current activity.

Activity 4-1   Measuring Northern Stars with Quadrant and Cross-staff

Answer Sheet   NAME _____

1. For the first set of measurements, record your results on the left side of the table below. Use the right side for the second set. Record your three measurements of each angle in columns #1, #2, and #3. After completing the measurements, calculate and record the averages and the errors. Remember not to start the second set of measurements until nearly two hours (and preferably longer) after your first set.

Date of your measurements: _____

	FIRST MEASUREMENTS					SECOND MEASUREMENTS				
	Time Started:					Time Started:				
	Time Ended:					Time Ended:				
	Measurement					Measurement				
ALTITUDES	#1	#2	#3	Ave.	Error	#1	#2	#3	Ave.	Error
Polaris	—	—	—	—	—	—	—	—	—	—
Eastern Star #1	—	—	—	—	—	—	—	—	—	—
Eastern Star #2	—	—	—	—	—	—	—	—	—	—
Western Star #1	—	—	—	—	—	—	—	—	—	—
Western Star #2	—	—	—	—	—	—	—	—	—	—
ANGLES FROM POLARIS	#1	#2	#3	Ave.	Error	#1	#2	#3	Ave.	Error
Eastern Star #1	—	—	—	—	—	—	—	—	—	—
Eastern Star #2	—	—	—	—	—	—	—	—	—	—
Western Star #1	—	—	—	—	—	—	—	—	—	—
Western Star #2	—	—	—	—	—	—	—	—	—	—
POSITION ANGLES	#1	#2	#3	Ave.	Error	#1	#2	#3	Ave.	Error
Eastern Star #1	—	—	—	—	—	—	—	—	—	—
Eastern Star #2	—	—	—	—	—	—	—	—	—	—
Western Star #1	—	—	—	—	—	—	—	—	—	—
Western Star #2	—	—	—	—	—	—	—	—	—	—

Part I
Discovering the Science of Astronomy

2. Use the graph paper below to plot the positions of each star you measured for both the first and second set of measurements. Label each position carefully.

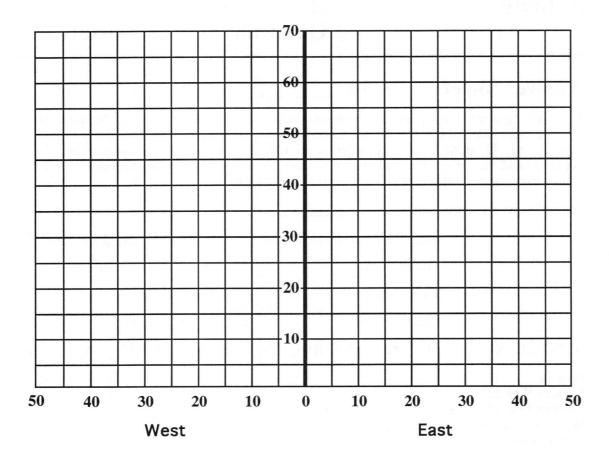

3. Answer Inquiries 4-1a on a separate piece of paper.

Kit Activity 4-2

██████████▌░░░░░░░

Mapping the Sky

After completing this activity, you should be able to do the following:

• Use a map of the full sky to locate the principal constellations and stars for your time and season.

• Use your quadrant and cross-staff to measure the positions and motions of stars in the sky, and interpret the measurements.

██████████▌░░░░░░░

Use the answer sheet at the end of the Activity.

On a clear, reasonably dark evening, go to an observing location with a clear horizon, as far from lights as possible. Locate the full-sky map that is appropriate to the time you are making your observations.

Find the most prominent features in each of the various directions in the sky. For example, in mid-September at 9 P.M., you would find overhead a large triangle consisting of the three bright stars Vega (in Lyra), Deneb (in Cygnus, the Swan), and Altair (in Aquila, the Eagle). These stars will have been prominent all summer and are known as the Summer Triangle. In the southeast, you would find Sagittarius, the Archer, looking like a teapot pouring hot tea on the tail of Scorpius, the Scorpion. In the west, you would find the bright star Arcturus in Boötes setting and, if conditions were good, Hercules as well. In the north, Cassiopeia will be high in the northeast, and the Big Dipper will be low in the northwest. Looking east, the most prominent feature will be the rising Great Square of Pegasus.

Once you have located some of these features, you can find others by starting at a known grouping and moving to an unknown one, using your map as a guide. For example, if you follow the handle of the Big Dipper from the bowl to the tail and continue on a curved line, you will easily be able to locate Arcturus (**Kit Figure 4-2-1**). "Follow the arc to Arcturus" is the saying. Following the arc, you can "speed to Spica." Similarly, such "star-hopping" beginning with the stars in the bowl of the Big Dipper can be used to locate Polaris, Vega, Capella, and Regulus.

In the winter, for example, in mid-January at 9 P.M., you would see Orion as a prominent constellation high in the southeast, with the brightest star in the sky, Sirius,

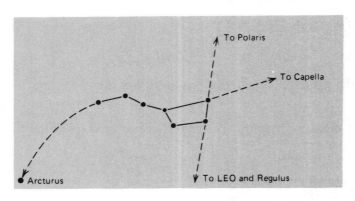

Kit Figure 4-2-1 Using the Big Dipper to locate other objects.

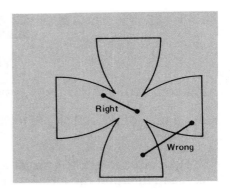

Kit Figure 4-2-2 Measuring angles on the star map. Be sure that your ruler does not cross a "cut" on the map.

just below it in the constellation of Canis Major. In the west you would find Pegasus, and Leo would be in the east. High overhead would be Taurus, with its bright star Aldebaran, and the Pleiades, a cluster of several hundred stars, of which seven can be seen under excellent conditions if your eyesight is keen.

Make the following measurements using your quadrant and cross-staff and be sure to record your results on the answer sheet.

1. Pick 3 non-circumpolar stars: one that is low in the east, another at medium altitude in the west, and a third one just to the east of the meridian. Measure their altitudes using your quadrant. Record the time of observation in your table. In September, you might use Pegasus in the east and Arcturus in the west; in January, Regulus in the east and Pegasus in the west.

2. Choose several bright stars (e.g., Vega, Deneb, and Altair, the Summer Triangle, in September; or Procyon, Sirius, and Betelgeuse, the Winter Triangle, in January). Measure the angle between each pair of stars using your cross-staff; record your data.

3. Pick a prominent constellation and sketch it on your answer sheet. In September, you could use Cygnus or Pegasus, for example; in January, Orion, Auriga, or Gemini might be good choices. Make multiple measurements of the angle between each pair of stars in the constellation with your cross-staff, recording the average angles on your sketch.

4. *At least* two hours after doing Step 1, repeat the measurements and record them. Keep track of your western star to make sure it does not set before you are able to remeasure it!

In making these measurements, do not forget to make at least three independent settings on each angle, average the results, and estimate the error of the measurement.

Having made your observations, you can check your angular measurements between pairs of stars by measuring the angular separation of the same stars on your cloverleaf star maps. If you use a millimeter scale, you will find that 1 millimeter on the star maps corresponds closely to 1° in the sky. For example, the distance on the map between Vega and Arcturus is almost 61 millimeters, and they are nearly 61° apart in the sky. (The correspondence is not perfect because there is some distortion due to the fact that the map is flat.) Be careful not to make any measurements that cross the splits on the map, as shown in **Kit Figure 4-2-2**, since such measurements would be inaccurate. Instead, choose another map on which the measurement can be made entirely on the map. For example, the angle between Antares and Spica cannot be measured on Map 8, but it can be measured on Map 6.

Kit Inquiry 4-2a According to your observations, in which part of the sky are the stars increasing their altitude? In which part of the sky are stars decreasing their altitude?

Kit Inquiry 4-2b Were there differences between your measurements of separations between stars in the sky and the same separation as measured on the map? What is the relationship between the separation and the amount of the difference?

Part I
Discovering the Science of Astronomy

Activity 4-2   Mapping the Sky

Answer Sheet   NAME _____

1. Record your first and second set of measurements on the left- and right-hand sides of the table below, using columns #1, #2, and #3 to record three independent measurements of the angles. Do not start the second set of measurements until at least two hours (and preferably longer) after the first set has elapsed.

Date of your measurements: _____

Altitudes of stars:

	FIRST MEASUREMENTS Time Started: Time Ended:					SECOND MEASUREMENTS Time Started: Time Ended:				
	Altitude Measurement					Altitude Measurement				
	#1	#2	#3	Ave.	Error	#1	#2	#3	Ave.	Error
Star low in east (name_____)	—	—	—	—	—	—	—	—	—	—
Star east of meridian (name_____)	—	—	—	—	—	—	—	—	—	—
Star medium in west (name_____)	—	—	—	—	—	—	—	—	—	—

Angles between stars:

	Angle Measurement between stars					Angle Measurement between stars				
Enter Star Names:	#1	#2	#3	Ave.	Error	#1	#2	#3	Ave.	Error
# 1 _____ to # 2 _____	—	—	—	—	—	—	—	—	—	—
# 1 _____ to # 2 _____	—	—	—	—	—	—	—	—	—	—
# 1 _____ to # 2 _____	—	—	—	—	—	—	—	—	—	—

(continued on next page)

2. Sketch a constellation indicate the values you measured for the angles between some of the stars. Next to each of these angles put the value for that same angle that you read off the star maps. Put the map values in a different color from your measurements.

3. Answer Kit Inquiries 4-2a, b below.

Kit Activity 4-3

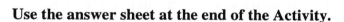

The Diurnal Motion of the Sun
Observed with a Gnomon

When you have completed this activity, you should be able to do the following:

• Set up a gnomon and use it to measure the altitude and azimuth of the Sun at various times of day.

• Define "azimuth."

• Determine the direction to true north from observations of the Sun.

• List and explain three effects that account for the different readings of sundials and ordinary clocks.

• Determine the time and azimuth of sunset from observations of the Sun.

Use the answer sheet at the end of the Activity.

Obtain a straight stick or rod that can be used for your gnomon. You might experiment with materials--a pencil stuck into the ground will work, or a piece of wire stuck upright in an eraser. Your gnomon should be about 6 inches long, although the exact length is not critical. It is important that the gnomon be *vertical;* you can use a carpenter's level or a weight on a string to assure this.

On a day that is clear enough to see shadows, find a location to set up your gnomon that will allow observations during most of the day. It should have a relatively clear western horizon and not be likely to be disturbed, since the gnomon cannot be moved once the observations have begun.

Now put a large piece of paper or cardboard under the gnomon and draw an approximate north - south line on it. Don't worry about being extremely accurate with the line, because the experiment itself will determine it correctly. If you can note the direction of the polestar at night and mark it on the ground, this will give you a usable line. A magnetic compass will be less accurate in most parts of the country, because it points toward magnetic north rather than true north. *Be sure the paper or cardboard is level and secured so the wind or foot traffic does not move it.*

Part I
Discovering the Science of Astronomy

At intervals during the day, mark the point on the paper where the end of the shadow falls, and write the time next to the mark. You should begin your observations before noon and make them regularly throughout the day, at least one per hour, and every 15 minutes when the Sun is near its maximum altitude. Continue making observations until as late in the afternoon as you can. Remember to note the date and time of your observations.

After making all your observations, draw a line from the point where the base of the gnomon stood to the points where the end of the shadow fell, and measure the length of each line. Also, measure the length of the gnomon. For each measurement, determine the altitude of the Sun as follows: First, compute the ratio

$$\frac{\text{length of stick}}{\text{length of shadow}}$$

and use **Kit Table 4-3-1** to find the angle corresponding to each ratio. (If you have a scientific calculator, you can also find the altitude using the arctan or tan^{-1} key with the above ratio as input.) For example, if the stick is 6 inches long and the shadow is 3 inches long, the ratio is 2 and the angle is about 63.5°. **Kit Figure 4-3-1** shows the geometry of the gnomon, shadow, and Sun. Record your data in the appropriate columns of the answer sheet for this activity. Finally, plot a graph of the altitude versus the time (see answer sheet, first graph). When you label the time axis on the graph, establish a scale that runs all the way to sunset. It is important to make sure that you use equally spaced time intervals on the graph. For example, if you let 5 marks represent an hour, continue this same scale across the axis. Answer the following questions and write your answers in the appropriate spaces on the answer sheet.

- **Kit Inquiry 4-3a** What was the greatest altitude of the Sun on the day you did your experiment?

- **Kit Inquiry 4-3b** From your graph, at what time did the greatest altitude of the Sun occur? Subjectively estimate the error in your determination of this time.

- **Kit Inquiry 4-3c** It is possible that you may not have been able to make your observations very close to sunset. However, you can estimate the time of sunset by extending the line on your graph until it intersects the "time" axis. Then read off the time of sunset at that point. Estimate the error of this extrapolation. Compare this time either with an observation you make of the time of sunset or with the time published in your local newspaper. If the times differ, suggest reasons for the difference.

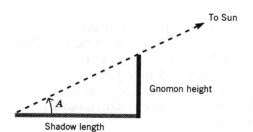

Kit Figure 4-3-1 The relationship between the altitude A of the Sun and length of the shadow.

Kit Table 4-3-1 Table of Solar Altitude Versus the Ratio of Stick Length to Shadow Length.

Ratio	Angle (°)	Ratio	Angle (°)	Ratio	Angle(°)
0.00	0	0.60	31	1.80	61
0.02	1	0.62	32	1.88	62
0.03	2	0.65	33	1.96	63
0.05	3	0.67	34	2.05	64
0.07	4	0.70	35	2.14	65
0.09	5	0.73	36	2.25	66
0.11	6	0.75	37	2.36	67
0.12	7	0.78	38	2.48	68
0.14	8	0.81	39	2.61	69
0.16	9	0.84	40	2.75	70
0.18	10	0.87	41	2.90	71
0.19	11	0.90	42	3.08	72
0.21	12	0.93	43	3.27	73
0.23	13	0.97	44	3.49	74
0.25	14	1.00	45	3.73	75
0.27	15	1.04	46	4.01	76
0.29	16	1.07	47	4.33	77
0.31	17	1.11	48	4.70	78
0.32	18	1.15	49	5.14	79
0.34	19	1.19	50	5.67	80
0.36	20	1.23	51	6.31	81
0.38	21	1.28	52	7.12	82
0.40	22	1.33	53	8.14	83
0.42	23	1.38	54	9.51	84
0.45	24	1.43	55	11.43	85
0.47	25	1.48	56	14.30	86
0.49	26	1.54	57	19.08	87
0.51	27	1.60	58	28.64	88
0.53	28	1.66	59	57.29	89
0.55	29	1.73	60		
0.58	30				

The next part of the activity is to determine the direction to true north. When the Sun is midway along its arc from the eastern to the western horizon, it will be located on the **meridian,** which is the circle that runs from the north point on the horizon, through the zenith, to the south point on the horizon[2]. When crossing the meridian, the Sun will be

[2]The meridian actually runs completely around the celestial sphere. At noon, the Sun crosses the *upper* meridian, while at midnight it crosses the *lower* meridian.

as high in the sky as it can be and, consequently, its shadow will be as short as it can be. For observers north of latitude 23.5°, when crossing the meridian the Sun will be *south* of the zenith point, and therefore its shadow will at that time be pointing to true north. **Kit Figure 4-3-2** shows this situation. Similarly, observers south of latitude -23.5° would find the shadow of the Sun pointing south, whereas in the tropics, whether the shadow points north or south depends on the time of year.

Take the sheet of paper on which you marked the Sun's shadow, draw a north - south line through the point where the gnomon was. You probably did not observe the exact moment when the shadow was shortest, so you will have to estimate where that occurred, as shown in **Kit Figure 4-3-3**.

Once you have marked the north - south line, you can measure the **azimuth** of the Sun at each observation. By definition, the azimuth is the angle that is measured clockwise from the north point on the horizon to the point on the horizon over which the Sun stands. Since the shadow of the Sun is on the *opposite* side of the gnomon from the Sun, this is the same as the angle, measured clockwise from the south to the shadow, as shown in **Kit Figure 4-3-4**.

Measure the azimuth of the Sun with a protractor and record your results. Then plot your results (see the second graph of the answer sheet).

• **Kit Inquiry 4-3d** On the basis of your observations, what was the azimuth angle of the Sun at its greatest altitude? What would you expect its azimuth to be then? Explain why your observations do or do not agree with what you expected.

• **Kit Inquiry 4-3e** Determine the azimuth of the Sun at sunset. Do this by extrapolating the azimuth graph to the moment of sunset determined earlier. The error of this estimate will be smaller the closer to sunset you were able to make observations. What is your estimate of the azimuth and its error? Record this on your graph.

• **Kit Inquiry 4-3f** Due west is azimuth 270°. Anything greater than this is north of west; anything less than this is south of west. Did the Sun set north or south of due west, or did it set exactly in the west? Record your answer on the graph. Does what you observed make sense for the time of the year you made your observation?

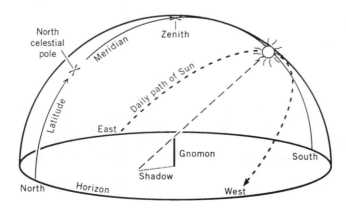

Kit Figure 4-3-2 The Sun at maximum altitude, north of latitude 23.5°.

Part I
Discovering the Science of Astronomy

You probably found that it was not *exactly* 12 noon when the Sun's altitude was greatest; in fact, it may have differed by over an hour. If we used ***apparent* solar time** in our daily life--the time read from a sundial--then noon would occur when the Sun is highest in the sky. However, civil clocks do not do this, for several practical reasons:

1. The Sun does not move at a uniform rate across the sky because of the noncircular orbit of the Earth around the Sun and other effects. The effect of this is that a sundial can be as much as 15 minutes fast or slow, compared to a uniformly advancing clock. The exact amount of the difference depends on the time of year.

2. The exact moment when the Sun is highest in the sky also depends on the observer's location within a time zone. For example, Kansas City, Missouri (KCMO), is a little bit east of Kansas City, Kansas (KCK); as a result, at any moment a sundial in KCMO will read a little later time than one in KCK. Since it would be inconvenient if every location had its own time, the convention of time zones has been adopted. All the occupants of a zone, roughly 15° wide in longitude, keep the same time on their clocks, equal to the average solar time at the central longitude of the zone.

• **Kit Inquiry 4-3g** What other effects would cause a difference between noon on the clock and the time of maximum altitude of the Sun?

• **Kit Inquiry 4-3h** Suppose there were a country where everyone set their clocks to 6 P.M. at the moment of sunset. What problems might arise as a result? (There is a country like this--Saudi Arabia.)

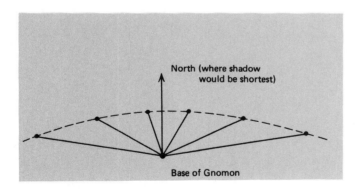

Kir Figure 4-3-3 Estimating the north-south line in the activity on the Sun's motion.

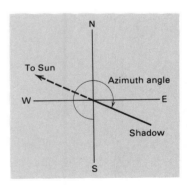

Kit Figure 4-3-4 Measuring the Sun's azimuth using the record of the Sun's shadow.

Part I
Discovering the Science of Astronomy

Activity 4-3   The Diurnal Motion of the Sun Observed with a Gnomon

Answer Sheet   NAME _____

Date of your Observations: _____

1. Record your observations of the Sun's altitude and azimuth in the table below:

Time	Shadow Length	Ratio	Altitude	Azimuth
____	____	____	____	____
____	____	____	____	____
____	____	____	____	____
____	____	____	____	____
____	____	____	____	____
____	____	____	____	____
____	____	____	____	____
____	____	____	____	____
____	____	____	____	____
____	____	____	____	____
____	____	____	____	____
____	____	____	____	____
____	____	____	____	____
____	____	____	____	____
____	____	____	____	____
____	____	____	____	____
____	____	____	____	____
____	____	____	____	____
____	____	____	____	____

2. On the graph, plot the altitude of the Sun versus time, for each observation.

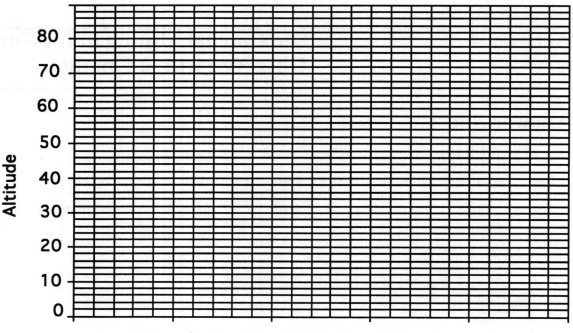

3. Inquiry 4-3a

4. Inquiry 4-3b

5. Inquiry 4-3c

Time of sunset from graph: _____ ± _____

Time of sunset observed directly or from newspaper: _____

6. Plot the azimuth of the Sun versus time on this graph.

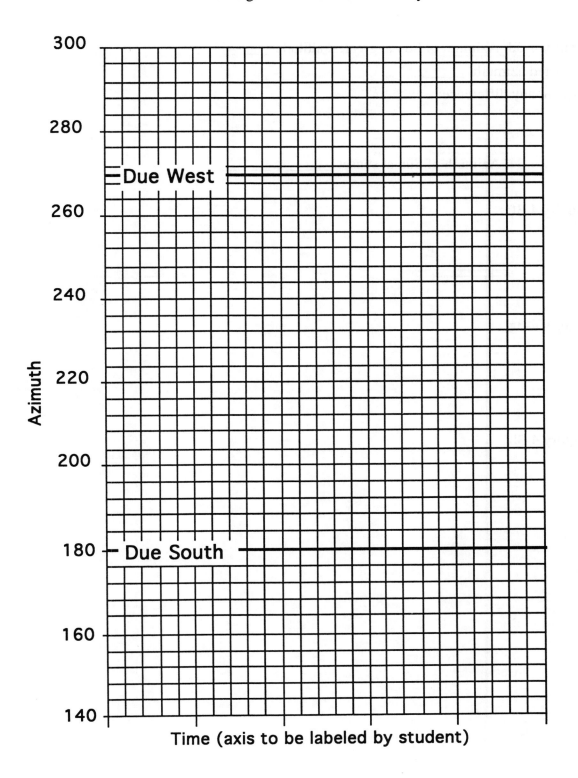

Inquiry 4-3d: Azimuth of Sun at greatest altitude _____

Expected azimuth _____

Explain any difference:

Inquiry 4-3e: Azimuth of Sun at sunset _____ ± _____

Inquiry 4-3f:

Inquiry 4-3g:

Inquiry 4-3h:

Kit Activity 4-4

The Motion of the Sun
with Respect to the Stars

When you have completed this activity, you should be able to do the following:

• Describe the apparent motion of the Sun among the stars.

Use the answer sheet at the end of the Activity.

By now you are probably familiar with the stars visible in the early evening. For this activity, you will need to choose a bright, easily recognizable star in the eastern part of the sky and another in the western part of the sky. For example, in the autumn you might choose Deneb (in Cygnus) in the east, and Arcturus (in Boötes) in the west. In the spring, Regulus (in Leo) might be a good choice for an eastern star and Aldebaran (in Taurus) in the west.

At least once every two weeks, *at the same time each night*, find the two stars you have chosen and measure their altitudes with a quadrant or your fist if necessary. Record the date and the altitudes on the answer sheet provided. Continue for at least six weeks to complete three sets of observations. Then answer the inquiries.

• **Kit Inquiry 4-4a** Does the star in the east get higher or lower as time progresses? By roughly how many degrees per month? What about the star in the west?

• **Kit Inquiry 4-4b** Since you made your observations at the same time of evening, the Sun will be roughly in the same position relative to the horizon each time (although not exactly, as will be found by those who do Kit Activity 4-5). Making this assumption, in what direction does the Sun appear to move among the stars during the course of the year (in an easterly or westerly direction)?

Activity 4-4   The Motion of the Sun with Respect to the Stars

Answer Sheet   NAME _____

NOTE: Be sure to make your observations at the same time each night.

Date	Star 1 Name _____ #1	#2	#3	Ave.	Error	Star 2 name _____ #1	#2	#3	Ave.	Error
_____	—	—	—	—	—	—	—	—	—	—
_____	—	—	—	—	—	—	—	—	—	—
_____	—	—	—	—	—	—	—	—	—	—
_____	—	—	—	—	—	—	—	—	—	—
_____	—	—	—	—	—	—	—	—	—	—
_____	—	—	—	—	—	—	—	—	—	—
_____	—	—	—	—	—	—	—	—	—	—
_____	—	—	—	—	—	—	—	—	—	—
_____	—	—	—	—	—	—	—	—	—	—

Inquiry 4-4a:

Inquiry 4-4b:

Kit Activity 4-5

Observing the Sunset Point

When you have completed this activity you should be able to do the following:

• Discuss how the location of the sunrise and sunset points change along the horizon throughout the year.

• Discuss how the location of the midday Sun on the meridian changes throughout the year.

The northward or southward motion of the Sun on the celestial sphere is easily detected in observations of sunrise or sunset. Choose an observing location with a clear western horizon (or eastern horizon, if you prefer to observe sunrises) and draw a sketch of it, including such obvious landmarks as trees, buildings, and telephone poles (see **Kit Figure 4-5-1** for a sample). A photograph works well, too. Find out the time of sunset from the newspaper or your instructor, or by observation. Just as the Sun is setting, mark on your drawing or photograph the point where it goes below the horizon. With your fist, or a cross-staff, measure the location of the sunset point with respect to some convenient object on the horizon chosen as a reference point (e.g., the telephone pole in Figure 4-5-1). All your later observations must be done from exactly the same location.

Measure this angle

Kit Figure 4-5-1 Measuring the angle between the Sun and a landmark. Be sure always to stand on the same spot.

After at least two weeks have passed, return to the same observing location to measure and draw the location of the sunset point. After two more weeks have passed, make a third and final observation, then answer the following questions. (To learn even more, continue your observations throughout the semester or even the entire year.)

• **Kit Inquiry 4-5a** Has the Sun moved northward or southward along the horizon?

• **Kit Inquiry 4-5b** Compute the rate of motion of the Sun in degrees per day by evaluating the ratio

$$\frac{\text{total change in the angle between the Sun and the reference point}}{\text{total number of days between the first and last observation}}.$$

Did the rate of motion change over the course of your observations?

• **Kit Inquiry 4-5c** From the results that you obtained, would you expect the midday Sun at the end of your observation period to be higher or lower in the sky than it was at the beginning? Explain your conclusion.

This activity is simple enough that you can construct your own answer sheet for it. If you have access to a celestial globe, this is an interesting set of observations to reproduce on that instrument.

Kit Activity 4-6

The Motion and Phases of the Moon

When you have completed this activity, you should be able to do the following:

• Describe the position and motions of the Moon relative to the Sun and stars.

• Explain the Moon's phase variations, correctly relating them to the Moon's position relative to the Sun.

• Explain why the sidereal and synodic months have different lengths.

Use the answer sheet at the end of the Activity.

For this activity, you will need to observe with your cross-staff *every 2 or 3 days for a couple of weeks* (try to take advantage of every clear night). Because each observing session should take only about 5 minutes, the total observing time is not large. At the end of a month, you will need to make one additional observation.

It is most convenient to start your observations just after new Moon when the Moon can conveniently be found low in the western sky just after sunset. The following directions assume that you are starting at this time.[3] You can find when the Moon will be at its new phase from a calendar, local newspaper, or public library.

Each observation, which should be made around sunset, consists of three parts: (a) Estimating the angle between the Sun and the Moon; (b) plotting the position of the Moon on a star map; and (c) making a sketch of the Moon's face to show its phase and the principal markings. The answer sheet for this activity contains a map on which the Moon's position can be plotted.

First, estimate the angle between the Sun and the Moon. If they are close, you can use your cross-staff. Alternatively, you can use the fist-and-hand method, or stand with one arm pointing toward the sunset point and the other arm pointing toward the Moon. The angle between your two arms will then be the angle between the Sun and Moon. As in **Kit Figure 4-6-1**, estimate this angle to the nearest 10° and record it on your answer

---

[3]You can start at other times during the lunar cycle; however, this has some disadvantages. For example, the Moon and Sun may be far from each other in the sky, making it more difficult to measure the angle between them. Also, near full Moon the sky is brighter and fewer stars can be seen to help determine the position of the Moon.

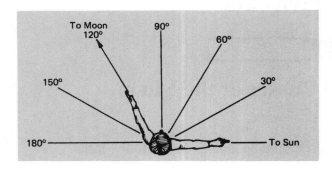

Kit Figure 4-6-1 Estimating the angle between the Moon and the Sun. In the figure, an angle of 120° is being estimated.

sheet. It is not important to be any more accurate than this. Note that it need not be dark for you to make this observation.

If you are making your observations after sunset, you should estimate the angle between the Moon and the point where the Sun *set,* and then add 15° for each hour past sunset. An advantage of observing an hour or two after sunset is that you will not have to wait for the sky to darken to locate the Moon with respect to the background stars.

Next, make a naked-eye sketch of the Moon's face. Draw the **terminator** (the line separating the bright and dark parts of the Moon) as accurately as you can and sketch the markings on the Moon's face as best as you can.

Finally, mark the Moon's location on the answer sheet's star map. If the Moon can be seen fairly close to recognizable stars, it will be sufficient to estimate its position by eye; on the other hand, if the Moon is very bright, or if there are no bright stars nearby, you will need to use your cross-staff. With your cross-staff, measure the angle between the Moon and two bright stars that you can locate on your map, as shown in **Kit Figure 4-6-2**. You should try to choose the stars so that the lines between them and the Moon meet at an angle, as shown. Then, plot the position of the Moon on your chart by drawing a circle around each star that has the same size as the angle you measured. *Use the vertical scale on the map to set your compasses to the right size.* The circles will intersect at two points, one of which is the position of the Moon. Choose the correct one (usually only one of them will be possible) and plot the Moon's position on the map at that point. If you are not sure which point to use, or if you want to double-check your measurement, pick a third star and draw the circle appropriate to it, after measuring the distance between it and the Moon. **Kit Figure 4-6-3** shows how to draw the circles.

As you proceed from night to night, you will find that the Moon changes position with respect to the stars, and that you will have to locate the Moon with respect to different stars. You can also use a bright planet instead of a star, provided that you can plot its correct position on the map.

After completing at least three observations of this kind, each separated by several days, answer the following questions.

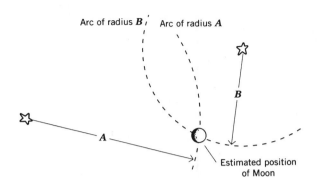

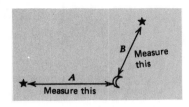

Kit Figure 4-6-2 Measuring the angle between the Moon and two bright stars.

Kit Figure 4-6-3 Plotting the Moon on your chart. The circles intersect at the position of the Moon.

- **Kit Inquiry 4-6a** In which direction does the Moon move, relative to the stars: from east to west, or from west to east?

You should find that the Moon moves approximately (but not exactly) along the ecliptic, which is represented as a horizontal line on the star map. The vertical axis is marked off in degrees on the map; by using this scale to measure the first and last observations of the Moon's position, you can determine how large an angle on the sky it has moved through. The rate of motion of the Moon is this number of degrees divided by the number of days between your first and last measurement.

- **Kit Inquiry 4-6b** Using the measurements you graphed of the Moon's motion relative to the background stars, what is the rate of motion in degrees per day? Approximately how long will it take the Moon to complete a 360° circuit of the sky with respect to the stars? (This time interval is called the **sidereal period.**)

- **Kit Inquiry 4-6c** Compare your Moon drawings with the estimates you have made of the angle between the Sun and the Moon on each date, and explain on your answer sheet how the Moon's phases come about. Use a diagram of the relative positions of the Earth, Sun, and Moon.

Observe the Moon once more when it has returned to just past new Moon again. Try to catch it when it returns to the same phase when you made your first observations. If weather prevents this, try to estimate from the previous observations when this would have occurred. (**Hint:** You know the rate of motion per day from your earlier observations.) The time taken by the Moon to return to the same *phase* is called the **synodic period**.

- **Kit Inquiry 4-6d** On the basis of your observations, how long is the Moon's synodic period?

Part I
Discovering the Science of Astronomy

| **Activity 4-6** | **The Motion and Phases of the Moon** |

Answer Sheet   NAME _____

1. Activities 4-6 and 4-7 both use the star charts found on the next two pages. These charts are centered on the ecliptic, the horizontal line across the center that shows the path of the Sun through the star during the year. On the ecliptic, the Sun's position is indicated for various dates separated by 10-day intervals. Follow the activity instructions to plot the positions of the moving bodies on these charts, using the degree scale on the vertical axis to measure your angles.

2. For Kit Activity 4-6, use the circles below as a guide to draw the appearance of the Moon as you observe it. Answer inquiries from the activity on a separate page and staple it to these pages.

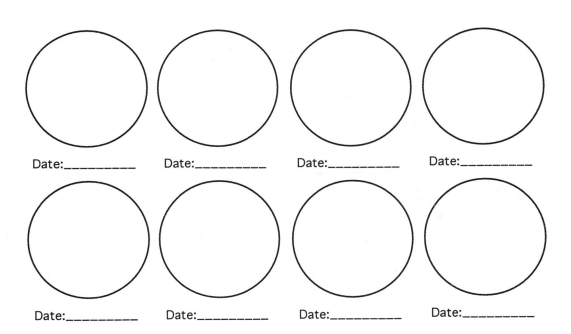

Date:_____   Date:_____   Date:_____   Date:_____

Date:_____   Date:_____   Date:_____   Date:_____

Part I
Discovering the Science of Astronomy

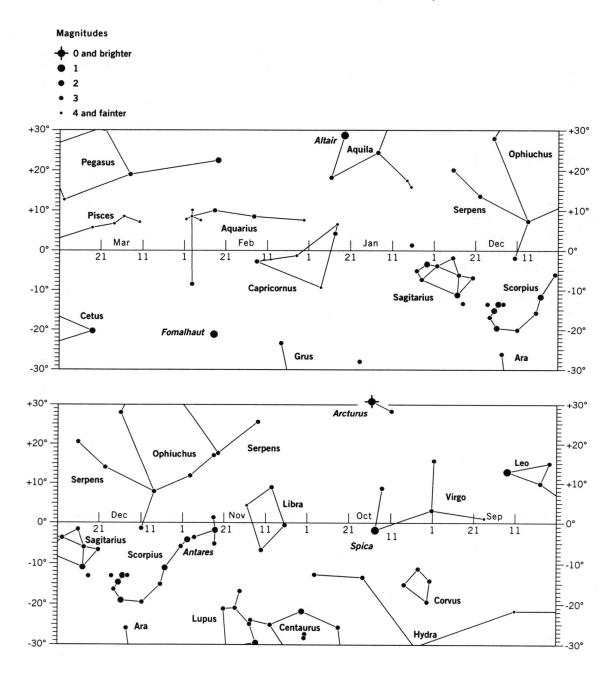

Magnitudes

● 0 and brighter
● 1
● 2
● 3
· 4 and fainter

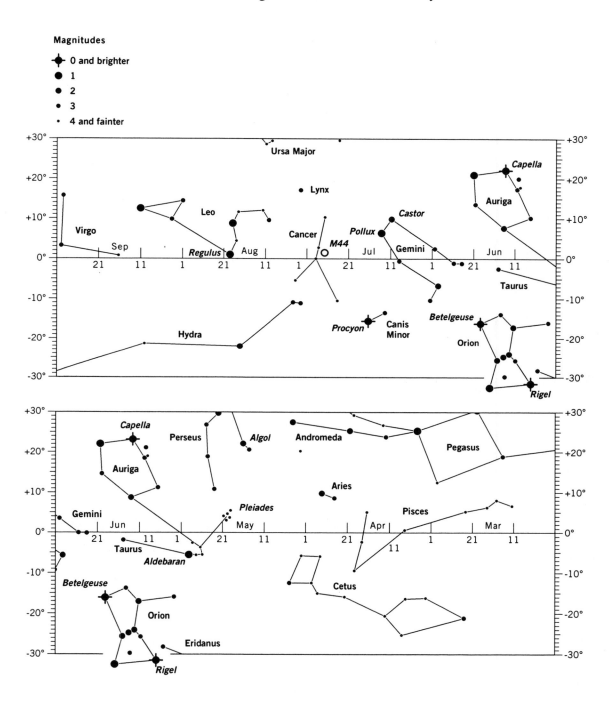

Part I
Discovering the Science of Astronomy

Kit Activity 4-7

The Motions of the Planets

When you have completed this activity, you should be able to do the following:

• Plot the position of any planet on your star map over a period of time.

• Describe in your own words the motions of the planets you observed.

• Define "retrograde motion" and recognize when it is taking place.

• Estimate directly from your observations the number of degrees per day that the planets you observed move.

Activity Table 4-7-1 gives the approximate locations, by constellation, of the planets visible to the naked eye for a 5-year period. Readers who have completed Activity Kit 4-6 can improve upon this approximate location by using the same techniques used there to locate the Moon. That is measure the angle between the planet and several nearby stars with the cross-staff, or make careful visual sightings of its position relative to nearby stars, then estimate its position on the map without using the cross-staff. If you wish, a pair of binoculars or a small telescope can be used to help locate the planet more precisely, since you will be able to see fainter stars. If you have a camera and tripod, a short time exposure will show the brighter stars and planets as short streaks on the film, and their relative positions can be estimated from this.

Whatever method you choose, you should follow the planets for a period of several months--the better part of a semester. Mars moves fairly rapidly, and you should be able to detect its motion with respect to the background stars in a few weeks; Jupiter and Saturn will probably take a month or more for the motion to become apparent. This depends on the care with which you make your observations. Whatever planet you choose, plot its position on your star map for as long a period of time as you can (at least a couple of months). Use the same star maps in Kit Activity 4-6.

Venus and Mercury, on the rare occasions when you can see Mercury, pose a somewhat different problem in that they never get very far away from the Sun in the sky. This makes it more difficult to measure their positions with respect to the stars, most of which are hard to see until the Sun is well below the horizon. Therefore, for these two planets you should measure their positions with respect to the Sun rather than the stars.

Activity Table 4-7-1 Locations of the Sun and Planets, 1995-1999[a].

DATE	Sun	Mercury	Venus	Mars	Jupiter	Saturn
Jan 1995	Sagittarius	Capricornus	Scorpius	Leo	Scorpius	Aquarius
Feb 1995	Capricornus	Capricornus	Sagittarius	Leo	Ophiuchus	Aquarius
Mar 1995	Pisces	Aquarius	Capricornus	Leo	Ophiuchus	Aquarius
Apr 1995	Pisces	Pisces	Pisces	Leo	Ophiuchus	Pisces
May 1995	Aries	Taurus	Pisces	Leo	Ophiuchus	Pisces
Jun 1995	Taurus	Taurus	Taurus	Leo	Ophiuchus	Pisces
Jul 1995	Gemini	Gemini	Gemini	Virgo	Scorpius	Pisces
Aug 1995	Leo	Leo	Leo	Virgo	Scorpius	Pisces
Sep 1995	Leo	Virgo	Virgo	Virgo	Scorpius	Pisces
Oct 1995	Virgo	Virgo	Virgo	Libra	Ophiuchus	Pisces
Nov 1995	Libra	Libra	Ophiuchus	Ophiuchus	Ophiuchus	Aquarius
Dec 1995	Ophiuchus	Sagittarius	Sagittarius	Sagittarius	Sagittarius	Pisces
Jan 1996	Sagittarius	Capricornus	Aquarius	Capricornus	Sagittarius	Pisces
Feb 1996	Capricornus	Capricornus	Pisces	Aquarius	Sagittarius	Pisces
Mar 1996	Pisces	Aquarius	Aries	Pisces	Sagittarius	Pisces
Apr 1996	Pisces	Aries	Taurus	Pisces	Sagittarius	Pisces
May 1996	Aries	Aries	Taurus	Aries	Sagittarius	Pisces
Jun 1996	Taurus	Taurus	Taurus	Taurus	Sagittarius	Pisces
Jul 1996	Gemini	Gemini	Taurus	Taurus	Sagittarius	Pisces
Aug 1996	Leo	Leo	Gemini	Gemini	Sagittarius	Pisces
Sep 1996	Leo	Virgo	Cancer	Cancer	Sagittarius	Pisces
Oct 1996	Virgo	Virgo	Leo	Leo	Sagittarius	Pisces
Nov 1996	Libra	Libra	Virgo	Leo	Sagittarius	Pisces
Dec 1996	Ophiuchus	Sagittarius	Libra	Leo	Sagittarius	Pisces
Jan 1997	Sagittarius	Sagittarius	Sagittarius	Virgo	Capricornus	Pisces
Feb 1997	Aquarius	Capricornus	Capricornus	Virgo	Capricornus	Pisces
Mar 1997	Pisces	Pisces	Pisces	Virgo	Capricornus	Pisces
Apr 1997	Pisces	Aries	Pisces	Leo	Capricornus	Pisces
May 1997	Aries	Aries	Taurus	Leo	Capricornus	Pisces
Jun 1997	Taurus	Taurus	Gemini	Virgo	Capricornus	Pisces
Jul 1997	Gemini	Cancer	Leo	Virgo	Capricornus	Pisces
Aug 1997	Leo	Leo	Virgo	Virgo	Capricornus	Pisces
Sep 1997	Leo	Leo	Virgo	Libra	Capricornus	Pisces
Oct 1997	Virgo	Virgo	Scorpius	Ophiuchus	Capricornus	Pisces
Nov 1997	Libra	Scorpius	Sagittarius	Sagittarius	Capricornus	Pisces
Dec 1997	Ophiuchus	Sagittarius	Capricornus	Sagittarius	Capricornus	Pisces
Jan 1998	Sagittarius	Sagittarius	Capricornus	Capricornus	Capricornus	Pisces
Feb 1998	Aquarius	Capricornus	Sagittarius	Aquarius	Aquarius	Pisces
Mar 1998	Pisces	Pisces	Capricornus	Pisces	Aquarius	Pisces
Apr 1998	Pisces	Pisces	Aquarius	Aries	Aquarius	Pisces
May 1998	Taurus	Aries	Pisces	Aries	Pisces	Pisces
Jun 1998	Taurus	Gemini	Aries	Taurus	Pisces	Aries

Part I
Discovering the Science of Astronomy

Activity Table 4-7-1 (continued)

DATE	Sun	Mercury	Venus	Mars	Jupiter	Saturn
Jul 1998	Gemini	Leo	Taurus	Gemini	Pisces	Aries
Aug 1998	Leo	Leo	Cancer	Cancer	Pisces	Aries
Sep 1998	Leo	Leo	Leo	Leo	Pisces	Aries
Oct 1998	Virgo	Virgo	Virgo	Leo	Pisces	Aries
Nov 1998	Libra	Ophiuchus	Virgo	Leo	Aquarius	Pisces
Dec 1998	Ophiuchus	Libra	Sagittarius	Virgo	Pisces	Pisces
Jan 1999	Sagittarius	Sagittarius	Capricornus	Virgo	Pisces	Pisces
Feb 1999	Aquarius	Aquarius	Pisces	Virgo	Pisces	Pisces
Mar 1999	Pisces	Pisces	Pisces	Libra	Pisces	Aries
Apr 1999	Pisces	Pisces	Taurus	Virgo	Pisces	Aries
May 1999	Aries	Aries	Gemini	Virgo	Pisces	Aries
Jun 1999	Taurus	Gemini	Cancer	Virgo	Pisces	Aries
Jul 1999	Gemini	Cancer	Leo	Virgo	Aries	Aries
Aug 1999	Leo	Cancer	Leo	Libra	Aries	Aries
Sep 1999	Leo	Virgo	Leo	Scorpius	Aries	Aries
Oct 1999	Virgo	Libra	Leo	Sagittarius	Aries	Aries
Nov 1999	Libra	Libra	Virgo	Sagittarius	Pisces	Aries
Dec 1999	Ophiuchus	Scorpius	Libra	Capricornus	Pisces	Aries

[a] The table gives the name of the constellation in which the Sun or planet is found. If the location is the same for the Sun and planet, they are probably too close to each other to view the planet.

Your data can be reported simply as a table of angles with respect to the Sun as a function of date.

Venus should be observed with your cross-staff every two to three days. You should measure the angle between Venus and the setting Sun. **Wait until the Sun has just set, so you can measure the angle between the setting point and the planet easily without endangering your eyesight.** The procedure for Mercury is the same, except that observations should be made every day over the several weeks that it is visible, since Mercury moves so rapidly.

After completing your observations, answer the following questions:

• **Kit Inquiry 4-7a** Did the planet you observed move east or west relative to the stellar background, or did it change its direction of motion? (NOTE: East is to the left on your map.)

• **Kit Inquiry 4-7b** If a planet moves from east to west, it is said to be in **retrograde motion**. Did you observe retrograde motion for any of the planets you observed? If so, which ones?

• **Kit Inquiry 4-7c** Roughly how many degrees per day did the planets you observed move, from your first observation to your last one?

Part I
Discovering the Science of Astronomy

Kit Activity 5-1

A Parallax Measurement

When you have completed this activity, you should be able to do the following:

• Describe what is meant by parallax.

• Measure the parallax of a nearby object with your cross-staff or fist, and determine its distance.

• Discuss the errors in making parallax measurements.

Use the answer sheet at the end of the Activity.

The object of this discovery is to measure the distance to a relatively nearby object by measuring its parallactic shift with respect to a much more distant object. The nearby object could be located a few meters away (in a room or down a hallway), or some tens of meters away (on a football field). You can use objects as much as a thousand meters away, if you have access to a safe rooftop with a clear view of the horizon and distant reference objects like buildings or radio towers. The distant reference object should be as far away as practicable--at least 10 times as far as the object whose distance is to be measured and preferably farther. **Kit Figure 5-1-1** shows how a parallax activity could be set up outside. The baseline AB should be from one-tenth to one-thirtieth of the distance to be measured. The nearby object illustrated is a person, and the background reference object is a pole.

To make your measurements, stand at point *A* and measure the angle between your nearby object and the distant reference object with your cross-staff or fist, as shown in **Kit Figure 5-1-2**. Repeat 3 to 5 times. Then move to the other end of your baseline (to point B) and measure the angle between your nearby object and the reference object again. From your two measurements, determine the total amount of the parallactic shift of the nearby objects relative to the reference object. For example, in Kit Figures 5-1-2 and 5-1-3, the parallactic shift is a total of 3°, the sum of the two measurements. Note, however, that if your foreground object and background object do not change sides when you move from one end of your baseline to the other end, then you should subtract the two angles rather than add them. Compute the distance to the nearby object from the

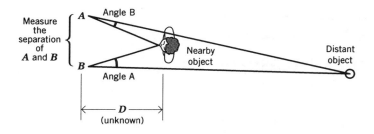

Kit Figure 5-1-1 The setup for
measuring parallax, as
seen from above.

angular size formula, rewritten as

$$D = 57.3° \times \frac{(\text{distance AB})}{\text{parallactic shift (in degrees)}}.$$

For example, if the baseline were 10 meters and the parallactic shift were 3°, the distance would be nearly 200 meters.

• **Kit Inquiry 5-1a** What did you obtain for the distance of the nearby object? How does this compare with the distance you obtained by pacing it off (or by measuring its distance on a map, if it is relatively distant)? Compute the percentage difference between your parallax determination and that you obtained through direct measurement.

• **Kit Inquiry 5-1b** Discuss the systematic and random errors in the two methods of determining the distance to the nearby objects. Which one do you think is more accurate and why?

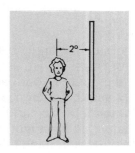

Kit Figure 5-1-2 Making the first measurement of the angle between the nearby object and the distant reference point.

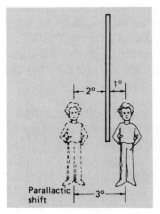

Kit Figure 5-1-3 Making the second measurement and determining the parallactic shift of the nearby object.

Activity 5-1   A Parallax Measurement

Answer Sheet   NAME _____

	POSITION A					**POSITION B**				
	#1	#2	#3	#4	#5	#1	#2	#3	#4	#5
Measurements:	___	___	___	___	___	___	___	___	___	___

Average: _____ ± _____   _____ ± _____

Parallactic shift: _____

Distance Computed: _____

Distance measured or paced: _____

Percentage difference between computed and measured: _____

Kit Inquiry 5-2b:

Part I
Discovering the Science of Astronomy

Kit Activity 8-1

Earthquakes

When you have completed this activity, you should be able to do the following:

• Given a seismograph record showing P- and S-waves, and a graph showing the arrival times for the waves as a function of distance, determine the location of a seismic epicenter.

Use the answer sheet at the end of the Activity.

This discovery activity requires a compass with a sharp pencil for drawing a circle.

An **epicenter** is the location on the Earth's surface above the place where an earthquake takes place. The actual location is called the **focus**. For this exercise, you will determine only the epicenter. Similar techniques may be used on any planetary surface on which a network of seismographs exists.

In **Kit Figure 8-1-1** are seismograms from three cities: Sitka, Alaska; Charlotte, North Carolina; and Honolulu, Hawaii. The first wave on each tracing is the P-wave and the second is the S-wave. The seismograms have all been adjusted to refer to the time in Charlotte.

• **Kit Inquiry 8-1a** At what times were the waves first detected at each location? Estimate the time to a tenth of a minute.

The time required for each type of wave to travel a certain distance through the Earth is shown in **Kit Figure 8-1-2**. The vertical distance between the curve labeled S and the curve labeled P is the difference in arrival times of the two waves and is an observed quantity. Once the delay in arrival time is observed, this graph can be used to determine the distance from the seismograph station to the epicenter. Use this travel-time curve to determine the distance from each seismograph station to the epicenter. Do this by placing a blank piece of paper along the time axis. Place a dot on the paper at the zero marking, and a dot at each of the S-minus-P time intervals you determined in the table above. Then, keeping the paper parallel to the vertical axis, slide the paper with the zero

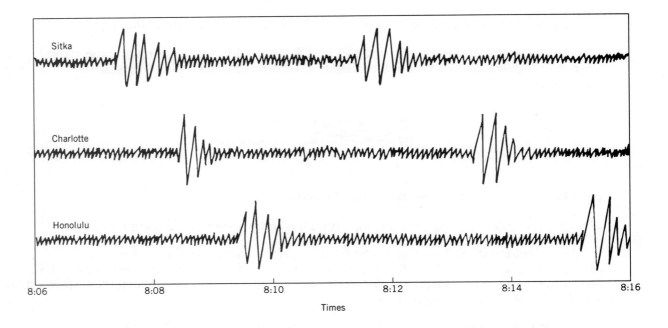

Kit Figure 8-1-1 Seismograms recorded at the city of Sitka, Charlotte, and Honolulu. The times shown have been adjusted to correspond to the time in Charlotte. (Reprinted with the permission of Macmillan Publishing Company from *Laboratory Manual in Physical Geology* 3/E by (American Geological Institute), Richard M. Busch. Copyright © 1990, 1993 by Macmillan college Publishing Company, Inc.)

dot always on the P curve until the dot for each city is on the S curve. You can then read off the distance to the epicenter by reading where the edge of the paper intersects the horizontal axis. Repeat for the other S-minus-P times.

• **Kit Inquiry 8-1b** What is the distance of the epicenter from each station?

Part II
Discovering the Nature and Evolution of the Solar System

To find the epicenter, you need to use the world map in **Kit Figure 8-1-3**. First, locate and mark the location of each station using the latitude and longitude as follows:

	Latitude	Longitude
Sitka	57° N	135° W
Charlotte	35° N	81° W
Honolulu	21° N	158° W

Next, place the point of your compass at the appropriate city on the map, and draw a circle having a radius equal in distance to the epicenter. Repeat for each station. Ideally, all three circles intersect at a point, which is the epicenter. More than likely, they do not intersect at one point. However, there may be a small triangle, with curved sides formed by pieces of the circles. The center of this triangle is the location of the epicenter. (If there is no such triangle, estimate the best location from the graph.)

• **Kit Inquiry 8-1c** What is the longitude and latitude of the epicenter?

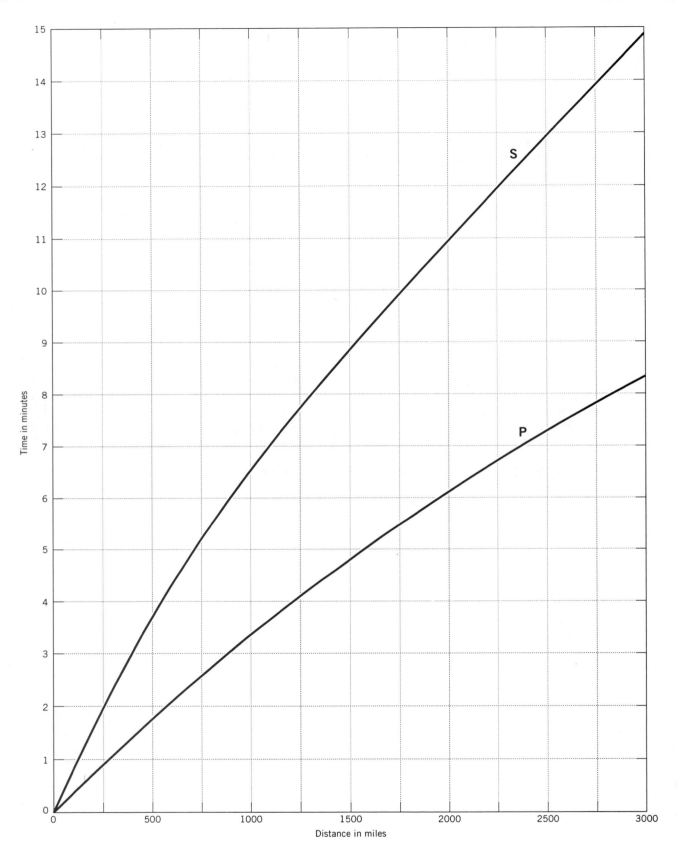

Kit Figure 8-1-2. Travel-time curves for the P- and S-waves through the Earth. (Reprinted with the permission of Macmillan Publishing Company from *Laboratory Manual in Physical Geology* 3/E by (American Geological Institute), Richard M. Busch. Copyright © 1990, 1993 by Macmillan college Publishing Company, Inc.)

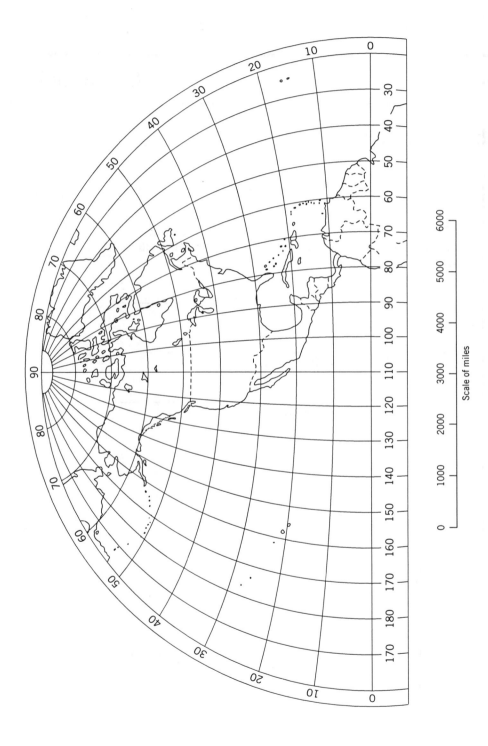

Kit Figure 8-1-3. Map of the world for plotting seismic results. (Reprinted with the permission of Macmillan Publishing Company from *Laboratory Manual in Physical Geology* 3/E by (American Geological Institute), Richard M. Busch. copyright © 1990, 1993 by Macmillan college Publishing Company, Inc.)

Activity 8-1   Earthquakes

Answer Sheet   NAME _____

Kit Inquiry 8-1a:

P-Wave arrival time	S-Wave arrival time	S minus P (minutes)
_____	_____ **Sitka**	_____
_____	_____ **Charlotte**	_____
_____	_____ **Honolulu**	_____

Kit Inquiry 8-1b:

	Distance (miles)
Sitka	_____
Charlotte	_____
Honolulu	_____

Kit Inquiry 8-1c:

The epicenter is located at a latitude of _____ and a longitude of _____.

Kit Activity 8-2

Surface Details from
Lunar Orbiter Photographs

When you have completed this activity, you should be able to do the following:

• Interpret Lunar Orbiter photographs of the Moon in terms of the kinds of features visible.

• Evaluate the evidence surface features supply for various processes that have taken place on the Moon.

• Deduce the sequence of ages for a set of features in a region of the Moon based on an examination of Lunar Orbiter photographs.

Kit Figures 8-2-1, **8-2-2**, and **8-2-3** show three of the many photographs transmitted to Earth by the Lunar Orbiter spacecraft. Each is provided with a grid for locating particular features.

Kit Figure 8-2-1 shows an area of approximately 30 km by 40 km located on the lunar equator just west of the center of the disk. On the original photograph, the smallest craters that can be seen are about 30 meters in diameter.

• **Kit Inquiry 8-2a** Compare the features marked 1 and 2 on this photograph. Which is raised and which is depressed, and from what direction is the sunlight coming? How did you come to your conclusions? (Don't just say that it "looked" a certain way. Present an answer that you could use to convince someone who had a different conclusion. For example, suppose you decided that feature 1 is raised and that the light came from right. How would you convince someone who claimed that feature 1 was depressed and that the light came from the left?)

• **Kit Inquiry 8-2b** Some obvious features in this photograph are:

(a) *Domes*--moundlike raised features of low elevation.

(b) *Mountains*--elevated features with steep slopes.

(c) *Craters*--circular depressions, rather regular in form.

Locate and give the map coordinates (e.g., H-6) of at least one good example of each of the above.

Kit Figure 8-2-2 shows a region near the lunar equator almost 200 km on a sidewhere both mountains and maria are easily seen. Referring to this figure, answer the following questions.

• **Kit Inquiry 8-2c** What evidence can you find in this figure that at some time in the past the material of the mare regions flowed across the landscape?

• **Kit Inquiry 8-2d** What features on the photograph have clearly been formed *after* this material stopped flowing?

• **Kit Inquiry 8-2e** Find features in Kit Figure 8-2-2 that could be described as isolated peaks. Also, find an example of a crater that has been submerged beneath flowing mare material.

• **Kit Inquiry 8-2f** Find an example of a crater that is superimposed on another crater. How can such a superposition be interpreted?

The superposition of one feature on another can be used to deduce a time sequence of events. For example, if one crater is superimposed on a second, it is clear that the crater underneath is the older one. This is just one example of the use of the principle of superposition. In Kit Figure 8-2-3, use this principle to establish a sequence of the history of the region around the crater with the large central peak. If you are careful, you should be able to distinguish about five different epochs in the history of the region.

• **Kit Inquiry 8-2g** What would you judge to be the oldest feature in the region? The youngest? Justify your choices. (**Hint:** In addition to the principle of superposition, you can also make use of the fact that older features will probably appear more eroded and less distinct than younger ones.)

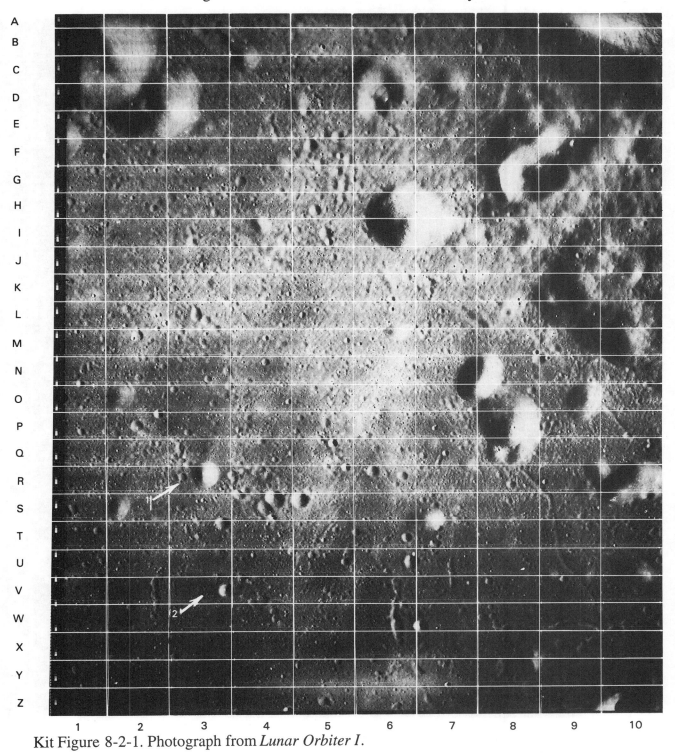

Kit Figure 8-2-1. Photograph from *Lunar Orbiter I*.

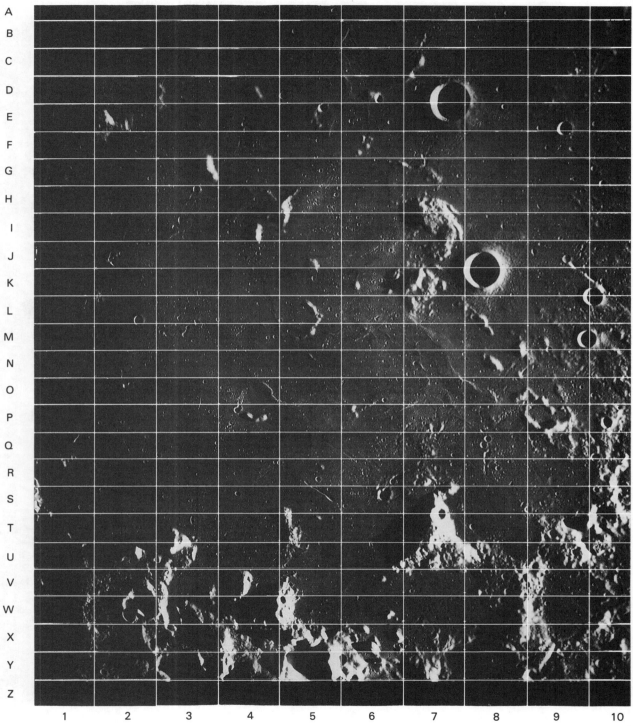

Kit Figure 8-2-2. Photograph from *Lunar Orbiter I*.

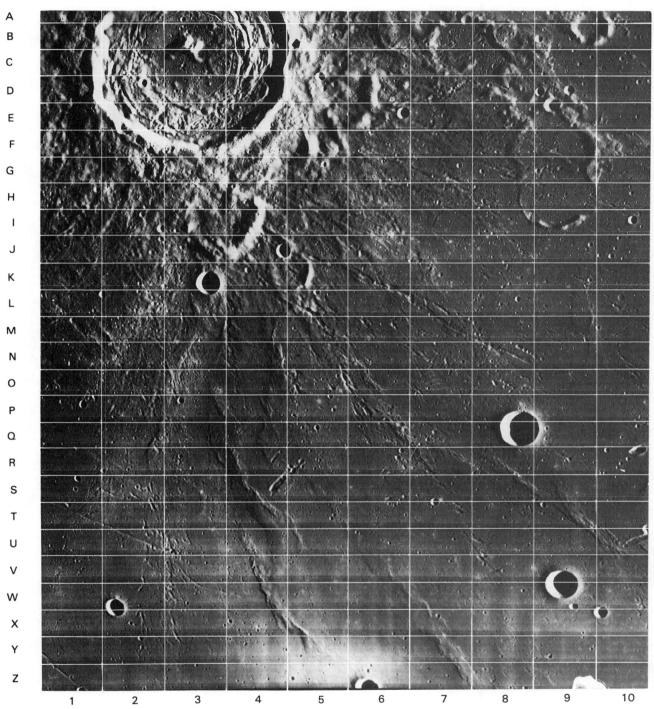

Kit Figure 8-2-3. Photograph from *Lunar Orbiter I*.

Part II
Discovering the Nature and Evolution of the Solar System

Kit Activity 10-1

[black and white bar graphic]

The Rotation of Jupiter

When you have completed this activity, you should be able to do the following:

• Given photographs of Jupiter taken at known times, determine its period of rotation.

[black and white bar graphic]

Use the answer sheet at the end of the Activity.

Kit Figure 10-1-1 contains two photographs of Jupiter in blue light taken 66 minutes apart on August 17, 1985. The Great Red Spot has clearly moved across the north/south line (the central meridian). By determining the fraction of a complete rotation through which the Great Red Spot has moved during these 66 minutes, you can then determine Jupiter's rotation period.

The angular distance through which the Great Red Spot has moved can be found approximately[4] from

Angular distance = 57.3° × $\dfrac{\text{measured linear distance through which it moved}}{\text{measured diameter at the latitude of the Red Spot}}$.

Use a ruler to measure the distance in millimeters through which the Great Red Spot has moved. Measure Jupiter's diameter at the latitude of the Spot.

• **Kit Inquiry 10-1a** Through what angular distance did the Red Spot move in the time interval?

• **Kit Inquiry 10-1b** Through what angle did it move per hour?

• **Kit Inquiry 10-1c** Given that in one complete rotation the Spot rotates through 360°, what is the period of rotation?

[4]For readers comfortable with trigonometry, the exact expression is given by

Sin (angle) = $\dfrac{\text{measured linear distance through which it moved}}{\text{measured diameter at the latitude of the Red Spot}}$.

Taken at
9:14

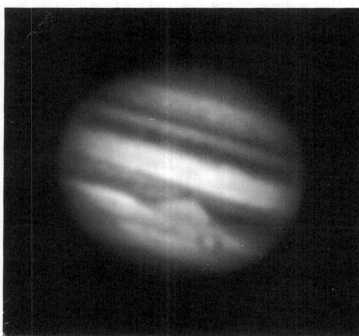

Taken at
10:20

Kit Figure 10-1-1. Jupiter photographed at Lowell Observatory
on August 17, 1985 at 9:14 and 10:20.

Activity 10-1   The Rotation of Jupiter

Answer Sheet   NAME _____

Inquiry 10-1a:

Angular distance moved between during time interval: _____

Inquiry 10-1b:

Angle per hour: _____

Inquiry 10-1c:

Period of rotation: _____

Part II
Discovering the Nature and Evolution of the Solar System

Kit Activity 10-2

The Galilean Satellites and the Mass of Jupiter

When you have completed this activity, you should be able to do the following:

• Given a drawing of the positions of the Jovian satellites on successive days, determine which satellite is which.

• Use the drawing, along with Newton's modification of Kepler's 3rd law, to determine the mass of Jupiter.

Use the answer sheet at the end of the Activity.

The Galilean satellites can be seen with a small telescope or a pair of binoculars. Were you to observe them every day for a couple of weeks, a drawing of their positions relative to the position of Jupiter would look like **Kit Figure 10-2-1**. The path defined by each satellite is defined by a smooth curve. The first thing you should do is to connect the successive points for *each* of the satellites (you might want to do it first on a photocopy of the figure).

• **Kit Inquiry 10-2a** On the basis of your data, what is the period of each of the satellites?

• **Kit Inquiry 10-2b** Measure the distance of each satellite in millimeters from the center of Jupiter. Express this distance in units of Jupiter's diameter by measuring its diameter in millimeters. Finally, express each satellite's distance from the center in kilometers.

• **Kit Inquiry 10-2c** Compute the mass of Jupiter. (Be careful to express your numbers in the proper units.)

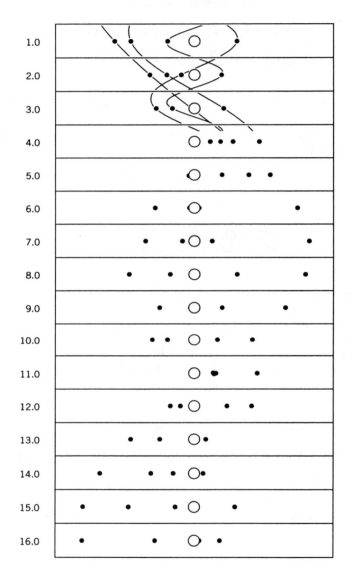

Kit Figure 10-2-1. The positions of Jupiter's satellites on successive days.

Activity 10-2   The Galilean Satellites and and the Mass of Jupiter

---

Answer Sheet   NAME _____

Inquiry 10-2a: Period from observations

Io   _____
Europa   _____
Ganymede   _____
Callisto   _____

Inquiry 10-2b:

Diameter of Jupiter measured on figure: _____ mm.

	Measured distance from center (mm)	Distance from center relative to Jupiter diameter	Distance from center (km)
Io	_____	_____	_____
Europa	_____	_____	_____
Ganymede	_____	_____	_____
Callisto	_____	_____	_____

Inquiry 10-2c:

The computer mass of Jupiter is _____.

Part II
Discovering the Nature and Evolution of the Solar System

Kit Activity 11-1

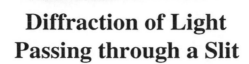

Diffraction of Light
Passing through a Slit

When you have completed this activity, you should be able to do the following:

• Define "diffraction" and explain the reasons for its occurrence.

• State how diffraction varies with the wavelength (color) of the light and the width of the aperture through which it travels.

In your kit you will find a cardboard pattern for an adjustable slit. Assemble it according to the following instructions. **(You will be using this slit again later on, so after this activity be sure to save it.)**

Construction of an Adjustable Slit

In the kit, the pieces that assemble to become an adjustable slit are the piece of black vinyl and the component pieces labeled PART A, PART B, and PART C.

Push out SLIT ASSEMBLY PART A and punch out the square hole. Push out PART B and staple it onto PART A where indicated. Take the black vinyl material and with a good pair of scissors cut it in half lengthwise. Place one piece of the black vinyl along the line indicated on PART A, as shown in **Kit Figure 11-1-1**, then glue, staple, or tape it into place. Make sure that the edge you cut is along the line.[5]

---

[5] If the black vinyl is not available, a double-edge razor blade, *carefully* broken in half lengthwise, makes an excellent slit for the spectrometer.

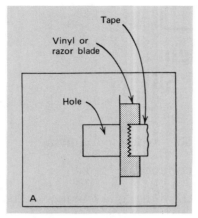

Kit Figure 11-1-1 Beginning the assembly of the adjustable slit.

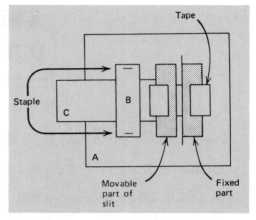

Kit Figure 11-1-2 The completed slit assembly.

Take the other half of the vinyl and glue, tape, or staple it to the movable slide PART C, with the cut edge facing to the right.

Now insert PART C under PART B to complete a movable slit assembly, as indicated in **Kit Figure 11-1-2**. Be sure that the sliding piece, PART C, fits snugly under PART B yet is still able to move freely.

Observations with the Slit

Find yourself a small, bright source of light, such as a distant street lamp at night, or a bare light bulb observed from a considerable distance away (30 feet or more). You may also use the image of the Sun *as reflected in a distant pane of glass or a convex surface such as chrome on a car;* **do not look directly at the Sun.** Hold the slit just before your eye, as shown in **Kit Figure 11-1-3**, and observe the source of light as you *slowly* close the slit. It takes a little practice to control the slit; it is probably easiest if you hold the slide between your thumb and forefinger. When the slit becomes very narrow, about the thickness of a sheet of paper, you will see the light begin to spread out. The more you close the slit, the more the light spreads out. This occurs because the path of the light is bent as it passes through the slit; that is, the slit *diffracts* the light.

• **Kit Inquiry 11-1a** Describe the appearance of the source of light at different slit widths. You may find it best to draw a sketch.

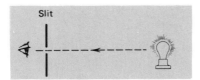

Kit Figure 11-1-3 Using the slit to observe diffraction.

• **Kit Inquiry 11-1b** Is the source spread out in a direction that is parallel to the *length* of the slit or perpendicular to it?

• **Kit 11-1c** Is the source spread out more when the slit is wide or when it is narrow?

Diffraction is both a burden and a blessing to the astronomer. It is a burden because whenever light passes through any aperture in an optical system, such as the lens of a telescope, it will be affected to some degree. The diffraction that occurs in a telescope, for example, tends to blur star images slightly and makes it impossible to resolve details of an image that are smaller than some angular size. On the other hand, since the amount of the diffraction depends on the wavelength of light, diffraction can be used to separate light into its component wavelengths for analysis.

Kit Inquiry 11-1d Since the lens of a telescope acts as an opening to diffract light, one would expect that the amount of blurring in star images would depend on the size of the lens, just as the amount of blurring you observed with your naked eye depended on the width of the slit. Which do you think would produce sharper (i.e., less blurred) images, a telescope with a large-diameter lens or one with a small lens?

If you narrow down your slit carefully, you should be able to see the spreading of the light in a direction perpendicular to the length of the slit. The spread-out light will consist of light and dark bands. If you see these dark bands, you are witnessing the phenomenon of the **interference** of light, another indication of light's wave nature.

Part III
Discovering the Techniques of Astronomy

Kit Activity 11-2

The Diffraction Grating

When you have completed this activity, you should be able to do the following:

• Describe the appearance of a variety of light sources when observed through the diffraction grating.

Your equipment packet contains a diffraction grating. It looks like a 2 x 2-inch photographic slide with clear plastic mounted in it; the lines are ruled on the plastic so finely that you cannot see them (there are on the order of 1000 lines per millimeter). Being careful not to touch the surface of the grating itself, hold the grating directly in front of your eye, as shown in **Kit Figure 11-2-1**. Examine a variety of light sources (at night)--street lamps, neon signs, fluorescent fixtures, incandescent bulbs--and see how the action of diffraction as a function of wavelength separates the light into its component colors. (If you do not immediately see a spectrum, rotate the grating 90 degrees.) Do you notice that different light sources emit their light in different patterns? Describe the differences in the appearance of the various light sources.

Kit Figure 11-2-1. Using the diffraction grating.

Part III
Discovering the Techniques of Astronomy

Kit Activity 11-3

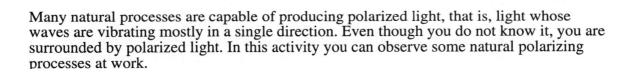

Polarized Light

When you have completed this activity, you should be able to do the following:

• Describe how you can produce and detect polarized light.

• Describe how you know the sky is polarized.

Many natural processes are capable of producing polarized light, that is, light whose waves are vibrating mostly in a single direction. Even though you do not know it, you are surrounded by polarized light. In this activity you can observe some natural polarizing processes at work.

Your instrument packet includes 2 pieces of polaroid filter. These filters are made from plastic that has been heated and stretched so that its pencil-like molecules are aligned with one another. When light waves pass through the filter, the aligned molecules absorb all the light waves *except* those with vibration components perpendicular to the aligned molecules. Normal, unpolarized light passing through such a filter thus becomes polarized.

Take one of the filters and look at a light bulb. Rotate the filter; do you observe any change in the appearance of the bulb as you rotate it? While looking at the light bulb, take the other (identical) piece of polaroid filter, place it on top of the first one, and look at the light bulb. Note what you see, if anything. Now, rotate one piece a quarter turn with respect to the other and again look at the light source through both filters.

• **Kit Inquiry 11-3a** Describe what you observed, and explain why the light is bright in one situation and dark in the other.

When the filters are parallel to each other, the light polarized by the first filter is able to pass through the second filter. However, when the filters are perpendicular to one another, the first filter transmits light in only one plane, which is perpendicular to the transmission direction in the second filter. For this reason, the light is blocked and no light transmission occurs. Now, continuously rotate one filter with respect to the other while observing the light.

• **Kit Inquiry 11-3b** Describe what you observed while rotating the filter.

Nature produces polarized light in various ways. To see this, look at light from a variety of sources--light reflected off the road, smooth and rough surfaces such as glass and concrete, and the like.

• **Kit Inquiry 11-3c** In your own words, describe what you see. Is the light in some circumstances more strongly polarized than in others?

(If you want to see some interesting effects, take some clear plastic or cellophane, and hold it *between* the pieces of polaroid filter while you rotate them.)

Lastly, take your filter outside on a sunny day when the sky is deep blue, free of clouds or haze. Look at various parts of the sky through the filter and determine whether the sky is polarized. You should see the blue of the sky darken and then brighten again as you do this because the light from the sky is polarized by our atmosphere. The polarization is most strong in the direction 90° from the sun.

Planetary atmospheres and surfaces polarize light, as do the outer atmospheres of certain types of stars. Some variable stars show polarized light because their light is scattered from dust in a shell around the star. The light of stars passing through dust in interstellar space is polarized. There are also galaxies that emit polarized radiation produced by electrons moving rapidly in a magnetic field.

Kit Activity 12-1

How Lenses Form Images

When you have completed this activity, you should be able to do the following:

• Define the term focal length.

• Show that images formed by lenses are inverted.

• Describe how image size is related to the focal length of the imaging lens.

• Distinguish between real and virtual images.

Use the answer sheet at the end of the Activity.

The lenses for this activity are in your activity kit. We recommend you assemble the telescope in the kit at this time using the instructions in Kit Activity 12-3. You may then attach one-half of the telescope assembly to one end of your meter stick with a rubber band. Use your cross-staff as a holder for your viewing screen, or, while less desirable, put the viewing screen on a wall and hold the lenses by hand if you prefer.

Place an ordinary incandescent or fluorescent lamp at one end of an otherwise darkened room; a fluorescent lamp will work best because of its length. Go to the other end of the room; with the viewing screen close to the lens, hold your lens-mounted meter stick so that the light from the lamp passes through the lens and onto the screen.

Slowly move the screen away from the lens. As you do so, you will notice that the spot of light on the screen becomes smaller and smaller. At some point it will focus into a clear and sharp image of the source of light. Measure the size of the image and the distance of the lens from the viewing screen. The larger the source is, the larger its image will be; this is why a fluorescent lamp works particularly well. **Kit Figure 12-1-1** shows the setup of this experiment.

Kit Figure 12-1-1 Setup for the study of image
formation by a lens. The lamp is imaged onto the
screen by the lens.

When a sharp image of a *distant* source of light is formed, the distance between the lens and the screen is called the focal length of the lens (see **Kit Figure 12-1-2**). The focal length is a property of the lens that is determined by the curvature of its surfaces and the kind of material from which the lens is made. The source of light should be infinitely far away from the lens for this measurement to be exact, but the error will be small as long as the source of light is many times more distant from the lens than the lens from the screen.

• **Kit Inquiry 12-1a** What are the focal lengths of the two lenses in your kit?

• **Kit Inquiry 12-1b** What was the image size of the light source as produced by each lens? What is the distance of each lens from the light source?

• **Kit Inquiry 12-1c** For each lens, divide the image size by the focal length. Divide the actual size of the object by the distance from the lens to the object. What relationship appears to hold between the two quotients?

• **Kit Inquiry 12-1d** Was the observed image erect or inverted? Was it reversed right-to-left?

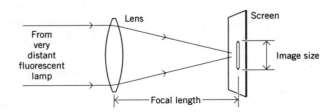

Kit Figure 12-1-2 Definition of the focal length
and image size.

STEREOSCOPIC VIEWS OF VENUS

Spacecraft orbiting distant planets provide planetary scientists with the capability of making 3-D images of features on a planet's surface. Such images provide a tremendously exciting new view of the planets. Three dimensional images allow planetary scientists to reconstruct the past history that produced what we observe. For example, these images allow scientists to see better the effects of faults and the collapse of surface features, as well as the lava flows of ancient volcanic events. We are pleased to provide users of *The Discovering Astronomy Activities Kit* with some incredible examples of this new tool.

The three images presented here were obtained with the *Magellan* spacecraft orbiting Venus. They were produced by analyzing radar signals reflected from Venus's surface. The resulting images resolve objects 120 meters in size, about the size of a football stadium. These images exaggerate the vertical scale by five to 10 times compared with the horizontal scale. You should view them not only from your normal reading distance, but also one to two arm-lengths away.

As you view these images, remember that you are looking at the surface of another world — a world that happens to be completely shrouded by clouds and with a surface that until recently was completely unknown to us. Try to imagine what processes might have produced what you are seeing. Try to imagine what it would be like to be on the landscape.

View the images with the blue filter over your right eye.

Plate 1. Hills and ridges stick out above a lava-flooded plain. Originally, the region was broken by crustal movement. Later, flows of volcanically produced matter covered the area, leaving only the highest peaks showing. The region in the image is 77 km across. Note the steepness of some of these mountains.

Plate 2. You are looking down the caldera of a volcano. The caldera, which is 50 km wide and 1.3 km deep, formed when the hot lava underneath the surface withdrew from its underground storage chamber and the surface collapsed into the chamber. How might you explain the flatness of the caldera bottom?

Plate 3. One of the largest impact craters on Venus, the 52 km wide Warren crater has a relatively smooth floor made of material of an unknown composition. The crater sits on top of a region that had been deformed by tectonic activity. Note the flatness of the crater's bottom; can you offer an explanation for it? How do you think the trench on the left of the crater was formed?

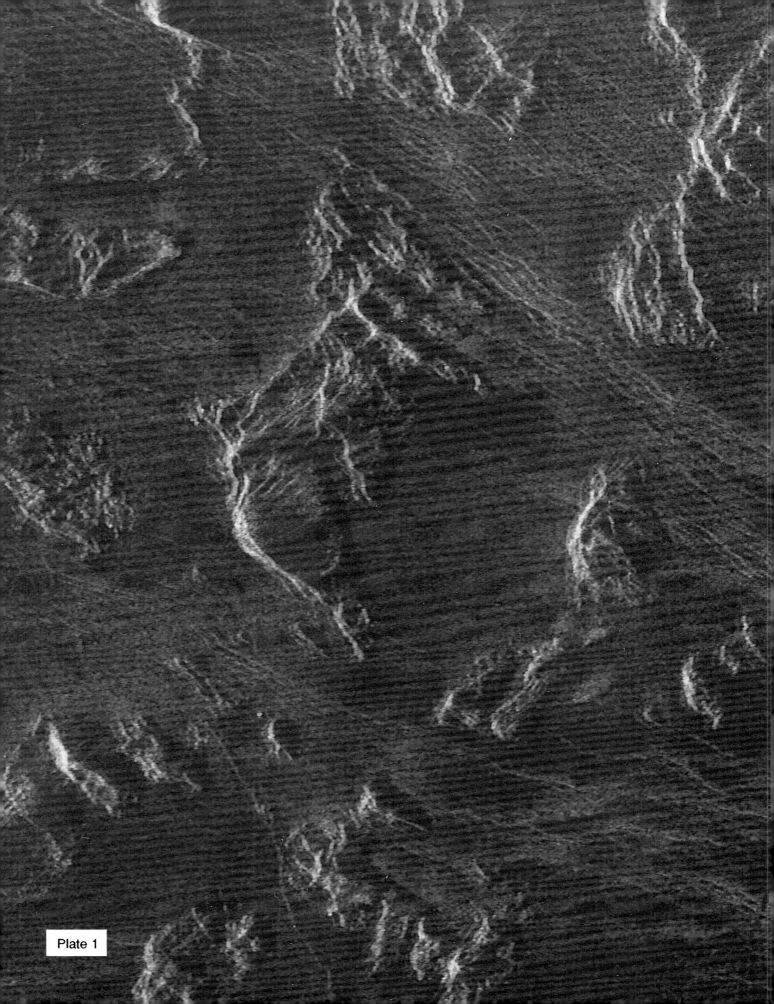

Plate 1

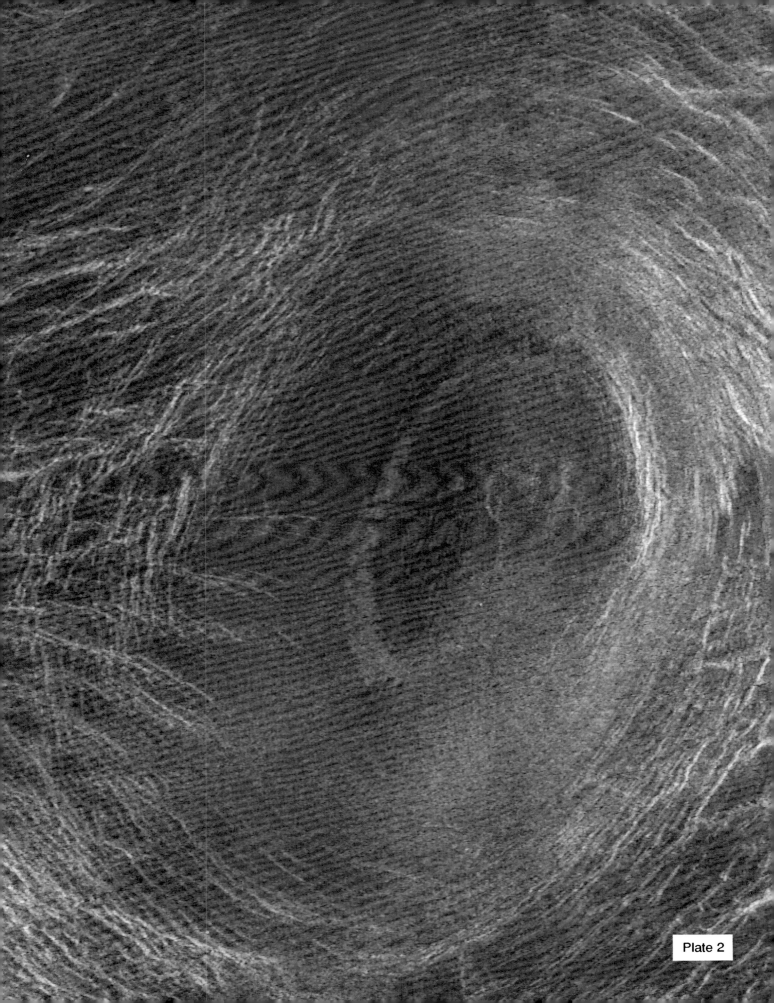

Plate 2

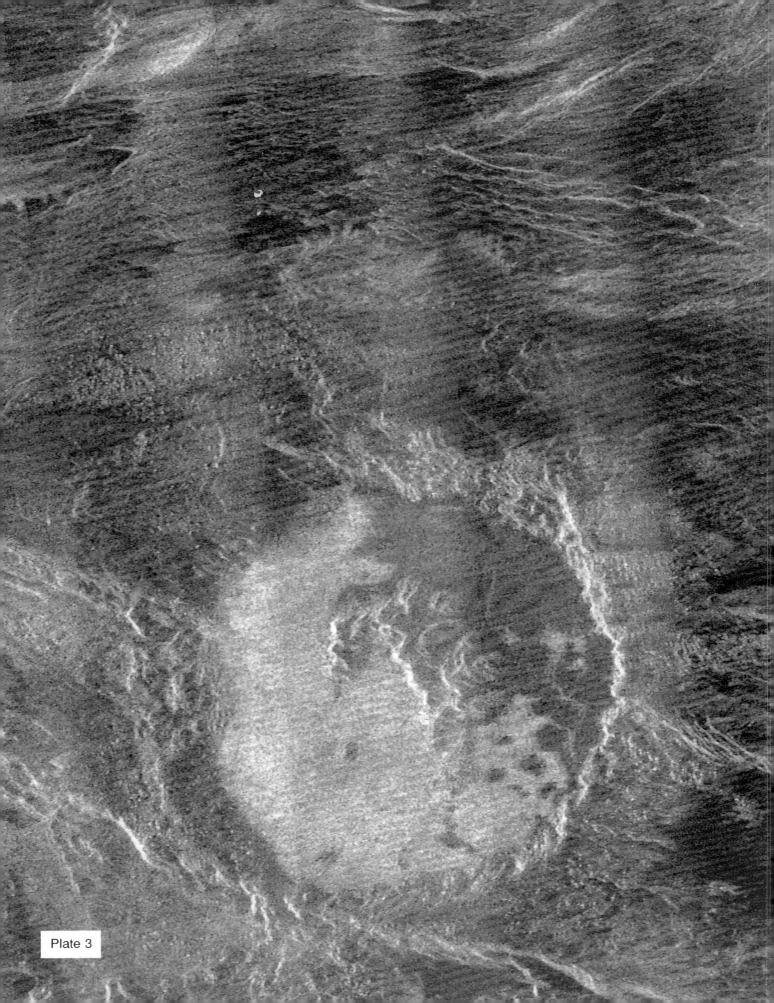

Plate 3

Part III
Discovering the Techniques of Astronomy

In this activity, you formed the image of an object on a screen. Images of this kind are called real images, to distinguish them from the virtual images as formed in a mirror. There is actually something physically present (or "real") at the focus position of a real image--a concentration of light energy. In a virtual image, however, no radiation passes through the image. A virtual image cannot be photographed by film placed where the image appears to be.

Kit Inquiry 12-1e What kind of image is formed in the pinhole camera?

Answers to Kit Inquiries

Kit Inquiry 12-1b. The lens with the longer focal length forms the larger image.

Kit Inquiry 12-1c. They are equal.

Kit Inquiry 12-1e. A real image.

Activity 12-1   How Lenses Form Images

Answer Sheet   NAME _____

Object (light source) size: _____ (mm)

	LENS A	LENS B
Focal length (mm):	_____	_____
Image size (mm):	_____	_____
Lens-to-object distance (mm):	_____	_____
Image size divided by focal length:	_____	_____
Object size divided by lens to object distance:	_____	_____

Inquiry 12-1d:

Inquiry 12-1e:

Kit Activity 12-2

Other Properties of Lenses

When you have completed this activity, you should be able to do the following:

• Describe what happens if half the lens is covered.

• Use a lens to magnify an object.

• Describe the type of image produced by a magnifying lens.

Take each lens in your kit and form a sharp image on the screen of the source of light you are using. With a piece of paper or your hand, cover up the top half of the lens.

• **Kit Inquiry 12-2a** What happens? Is the image "chopped in half"? Why or why not? (**Hint:** Draw a lens forming an image, and ask what would happen if half the lens were covered.)

A lens can be used as a magnifier. The shorter its focal length, the more powerfully it magnifies. Carefully remove the large lens from the telescope. Observe some small object by holding the lens just in front of your eye; move close enough to the object for it to be in good focus, as shown in **Kit Figure 12-2-1**.

Kit Inquiry 12-2b Does the magnified image appear closer to you or farther away than the actual object?

Kit Inquiry 12-2c What kind of image, real or virtual, do you observe?

Replace the lens in the telescope.

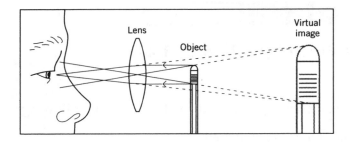

Kit Figure 12-2-1 Using a lens as a magnifier.

Answers to Kit Inquiries

Kit Inquiry 12-2a. The image will become dimmer all over. It is not chopped in half, because every part of the lens is capable of forming a complete image.

Kit Inquiry 12-2c. Virtual.

Kit Activity 12-3

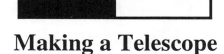

Making a Telescope

When you have completed this activity, you should be able to do the following:

• Compute the light gathering power, resolving power, and magnification of a telescope.

Use the answer sheet at the end of the Activity.

If you constructed the telescope earlier, skip the next two paragraphs. The remainder of the activity uses this telescope.

Your activity kit contains materials for making a small telescope that you can use to observe astronomical and terrestrial objects. You will find two pieces that fold out to become square tubes that fit into each other (see **Kit Figure 12-3-1**). The larger lens fits into the notch in the end flap of the tube marked A and then rotates 90 degrees to lock into place. The flap is then folded into the tube. (In other words, insert the lens into the hole so that one of the tabs is in the notch of the hole and the tab on the opposite side fits through the hole, so that the tab and the larger diameter of the lens form a channel for the flap. Push the lens into this channel and rotate it 90° to lock into place.) To assemble the eyepiece, place the other tube *with the letter B face down*. Place the small lens with the flat side down on a clean surface. Remove the backing from the self-stick lens holder, center its hole over the crown of the lens and press the sticky side down onto the lens until the lens adheres to it. Then stick the lens and its holder over the hole in flap B (the side opposite where the letter B is printed). There is an extra lens holder should you need it. If you have done this correctly, the flat side of both lenses will face your eye, which provides the best viewing.

To focus the telescope, aim it at some feature on the distant horizon (or the Moon) and slide one tube inside the other until the image is sharpest. Note that the image is inverted. If the inner tube wobbles inside the outer one, you can tighten it up by placing a folded piece of paper on one side of the inner tube. You might want to tape the two tubes together to hold the focus. Finally, attaching your telescope to a meter stick helps you aim the telescope better and hold it steadier.

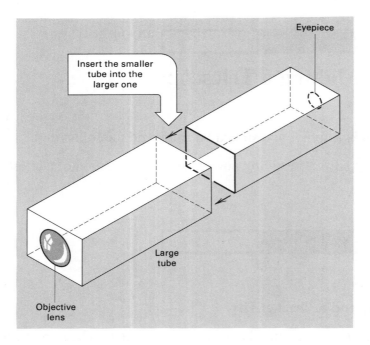

Kit Figure 12-3-1 The assembled telescope.

Kit Inquiry 12-3a Assuming that the dark-adapted eye has a diameter of one-fifth of an inch (0.5 cm), calculate how much more light-gathering power your telescope achieves.

Kit Inquiry 12-3b How does the resolving power of the objective of your telescope compare with your eye's resolving power?

Kit Inquiry 12-3c Compare the apparent angular diameter of the object with and without the telescope. About how many times does your telescope magnify?

You can do this estimation somewhat more precisely if you draw a set of parallel lines (on the board or a piece of paper) and look at those lines through the telescope with one eye, and without the telescope with the other eye. With practice, you can see both sets of lines simultaneously, superimposed in your mind's eye, and compare the scale directly. If you have trouble with this, it can help to look through a cardboard tube from a roll of paper towels.

Kit Inquiry 12-3d Calculate the magnification of your telescope using the focal lengths of the lenses that you measured earlier. Compare this value of the magnification with that found in Kit Inquiry 12-3c by finding the percentage difference between them.

Part III
Discovering the Techniques of Astronomy

Your small telescope is comparable to the first one that Galileo made for himself around the year 1609. You can use it for some simple astronomical observations.

Do not, under any circumstances, view the Sun with it or any optical system. Permanent, irreparable damage will occur to your retina.

You might also enjoy repeating Galileo's historical observations of the Moon, Venus and Jupiter with your instrument. You could vicariously share the excitement he felt in seeing for the first time such things as the satellites of Jupiter and features on the Moon.

Activity 12-3   Making a Telescope

Answer Sheet   NAME _____

(a)   Diameter of objective: _____

         Light gathering power: _____

(b)   Resolving power: _____

(c)   Apparent angular diameter

              with telescope: _____

              without telescope: _____

         Magnification: _____

(d)   Focal length lens A: _____

         Focal length: _____

         Magnification: _____

         Percentage difference: _____

Kit Activity 13-1

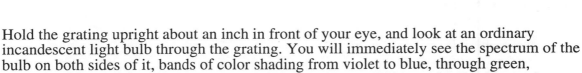

Observing Patterns
from Radiation Sources

When you have completed this activity, you should be able to do the following:

• Describe the spectra of an incandescent bulb and some common street lamps.

Hold the grating upright about an inch in front of your eye, and look at an ordinary incandescent light bulb through the grating. You will immediately see the spectrum of the bulb on both sides of it, bands of color shading from violet to blue, through green, yellow, and out to red. (If you do not see this pattern, rotate the grating 90 degrees.) Beyond the red you will see a second (fainter) spectrum like the first, and possibly a third. These "extra" spectra contain no information not found in the main spectrum, so we will not consider them here.

• **Kit Inquiry 13-1a** As you observe the spectrum of the incandescent bulb, do you see any regions between violet and red where *no* light is visible? That is, are there any gaps in the band of light?

A spectrum of the type emitted by an incandescent bulb is called a **continuous spectrum** because the bulb is emitting energy at *all* visible wavelengths, with no gaps. The radiation is emitted by a metal (tungsten alloy) filament in the bulb that is heated to a high temperature by passing an electric current through it.

A second type of spectrum can be seen by doing nighttime observations of the common bluish-white street lamps found in many locations. (If you are unable to find such a street lamp, use the yellow or pinkish ones found in many cities, or a neon light as a last resort.) Find such a lamp that is as far away from other interfering sources of light as possible and examine its light through the grating. (NOTE: To have enough contrast to see the spectrum you will have to do this observation after it is dark.) If you are looking at the right kind of lamp, you will see several images of the lamp in various colors: Blue, green, yellow, and probably red. But the light of the street lamp will *not* be spread into a continuous band; there are places in the spectrum where no image of the lamp is seen.

• **Kit Inquiry 13-1b** Which of the following *best* explains the appearance of the spectrum of a street lamp?

(a) The lamp is emitting energy uniformly at *all* visible wavelengths.

(b) The lamp is emitting energy at *most* visible wavelengths, but not at certain specific wavelengths.

(c) The lamp is emitting nearly all its energy at only a *few specific* wavelengths.

The spectrum emitted by such a street lamp could be called a **discrete emission** spectrum (energy is emitted only at certain wavelengths) to contrast it with the continuous spectrum of an incandescent lamp. When discrete emission appears, the radiation is being emitted by a **rarefied** (i.e. low density) transparent gas that acquires excess energy when electricity is passed through it.

Answers to Kit Inquiries

13-1a. No such regions should be visible.

13-1b. (c)

Kit Activity 13-2

Analyzing Light Sources
with a Spectrometer

When you have completed this activity, you should be able to do the following:

• Construct a simple spectrometer and calibrate it using the green line of mercury.

• Utilize the spectrometer to measure the wavelengths of the principal features in various common sources of radiation.

• Identify which of the three principal types of spectra is being observed from its appearance in the spectrometer.

• Identify common elements in radiation sources from their spectra, as observed in the spectrometer.

• Describe the principle by which a spectrometer operates.

Use the answer sheet at the back of this activity.

Part I. The Principle of the Spectrometer

A **spectrometer** is a device to view a spectrum using the eye. Its operation depends on a **diffraction grating** or a **prism** dispersing a beam of radiation into its component wavelengths or colors. **Kit Figure 13-2-1** shows the spectrometer whose parts are in the instrument kit. It illustrates the basic principles and components of the spectrometer. It makes use of the **slit** that you have already assembled and used; its purpose is to limit the light entering the instrument and to define better the resulting spectrum.

Consider what happens when you observe radiation of a single wavelength instead of the many wavelengths of an incandescent source. **Kit Figure 13-2-2** shows how the radiation coming through the slit falls on the grating and disperses through an angle that depends on the wavelength of the radiation. The radiation then travels toward the eye. It appears that the radiation comes from the direction of the scale; in other words, the eye

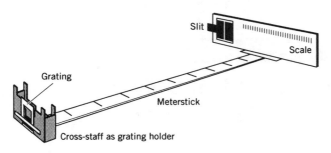

Kit Figure 13-2-1 The completed spectrometer.

sees a **virtual image** of the slit superimposed on the scale. The place where the virtual image of the slit is seen can be measured on the scale, giving a measurement of the wavelength of the radiation being observed.

If the radiation consists of many wavelengths, then the image of the slit will be seen in many different directions, one for each wavelength. In this way, the spectrum of a source containing any mixture of wavelengths is formed. If the spectrum is a discrete emission spectrum, bright lines will be seen at certain places on the scale and nothing will be seen elsewhere; if a continuous spectrum is being observed, radiation will appear to come from every position on the scale.

The spectrometer needs a slit to work properly. If the slit is too wide, then the virtual image of the slit will also be wide and the measurement of the wavelength will be less exact. Furthermore, if two lines in a discrete emission spectrum are very close together, as in the case of the yellow lines of sodium, then the images of the slit in the two wavelengths would become superimposed and only one line would be seen (**Kit Figure 13-2-3**). Thus, if the slit is too wide, the **resolving power**, or ability of the spectrometer to distinguish one spectral feature from another, would be reduced.

• **Kit Inquiry 13-2a** If the slit were 1 millimeter wide, about how far apart (in angstroms) would the wavelength of two lines have to be for the spectrometer to resolve them? (**Hint:** The virtual image of the lines would also be 1 millimeter wide. How many angstroms does 1 mm on the wavelength scale correspond to?)

Kit Figure 13-2-2 How the spectrometer works. A virtual image of the slit is seen in each color, superimposed on the scale.

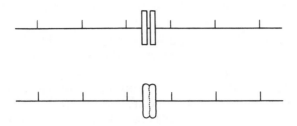

Kit Figure 13-2-3 The resolving power of the spectrometer
is dependent on the width of the slit. If the slit is too wide,
the observed spectral lines will merge and be indistinguishable.

Part II. Construction of the Instrument

To build your spectrometer, you will need a meterstick, the spectrum scale and its brace from the activity kit, the adjustable slit and the sliding piece of your cross-staff. Refer to Kit Figure 13-2-1 while you construct your spectrometer. Remove the scale from the piece of cardboard and remove the two rectangular pieces marked PUNCH OUT. Bend the scale along the indicated line. Slide the meterstick through the central hole so that the scale is toward the zero end of the stick. Set the scale perpendicularly to the stick so that it is located at exactly the 10 3/4-inch (27-cm) mark and secure it in place with tape. Push out the brace, staple it to the back of the scale, and tape it to the meterstick so that the scale remains perpendicular to the stick.

Now line up the adjustable slit with the mark on the scale and tape it into place. The adjusting tab should be at the left side of the scale. Finally, mount the *sliding* piece of your cross-staff in a backward position, with its perpendicular section exactly at the end of the meterstick, and secure it with tape. If you earlier made the cross-staff, you should have attached the grating holder at that time. If not, staple the grating holder from the instrument kit onto the sliding piece of the cross-staff to hold the grating. Place the grating into the grating holder. The completed spectrometer should resemble Kit Figure 13-2-1.

Part III. Observing with the Spectrometer

A. The Spectrum of an Incandescent Light Bulb

Be sure the grating is oriented so that the spectrum of a source is seen to the *side* rather than above and below the source. If the spectrum is above and below the source, rotate the grating. Adjust the slit so that it is about 0.5 millimeter (1/32 inch) wide. Stand in a semidarkened place, hold the meterstick up to your cheek, and point the entire instrument at an incandescent (ordinary) light bulb. You should be able to see the bulb through the grating *and* the slit, as shown in **Kit Figure 13-2-4**. (**Hint:** It will help in all experiments using the spectrometer if you can rest the center of the meterstick on a firm support.)

Move the slit horizontally back and forth until you see a streak of light (the spectrum) against the scale of the spectrometer. You may need to adjust the position of the grating slightly to get the spectrum properly superimposed on the scale. When the grating is adjusted properly, you should be able to read off the wavelengths of the various colors easily. Note that you can see the spectrum by merely holding the grating in front of your eye, as you did above. The purpose of the spectrometer is to get the dispersed light to land on the scale of the instrument, so that wavelengths of spectral features can be determined.

• **Kit Inquiry 13-2b** Describe in words the spectrum of an incandescent bulb. Over what range of wavelengths are the following colors found: Violet? Blue? Green? Yellow? Orange? Red? (NOTE: Eyes differ in their sensitivity to various colors, so do not be disturbed if your results do not agree with someone else's, or even if you cannot distinguish all the gradations listed here. An occasional reader may find for the first time that he or she is color-blind.)

B. Observing a Street Lamp

Now examine the emission from one of the bluish-white street lamps that are so common today. (This needs to be done at night to get the contrast necessary to see the spectrum.) Keep the slit narrow and line the instrument up as before. You should see a spectrum consisting of a few bright vertical "lines," energy emitted only at specific wavelengths. The dark spaces in between are wavelengths at which little, if any, energy is emitted. The line-like appearance of the spectrum is the result of light being forced to enter the instrument through the narrow slit. Because astronomers usually use a slit when taking spectra of celestial objects, the emission usually appears in the form of thin lines. As a result, a discrete emission spectrum in astronomy is often called a **bright-line spectrum.**

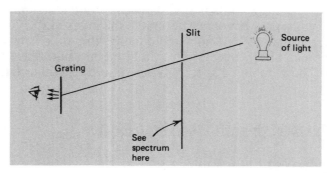

Kit Figure 13-2-4 Using the spectrometer. Light enters
the slit and the spectrum is seen superimposed on the scale.

Without the slit, the spectral features would still appear, but they would be "fatter" and it would be more difficult to measure their wavelengths precisely. Thus, the function of the slit is to shape the incoming light and thereby increase the amount of detail seen in the spectrum. (A photograph of the flash spectrum obtained during a total eclipse of the Sun provides an example of curved spectral lines.)

The bluish-white street lamp you are observing is filled with mercury vapor, which glows because of an electric current passing through it. As a check on the accuracy of your spectrometer, you should see a green emission line at a wavelength of 5461 angstroms (Å). If you see it at some other wavelength, you can adjust your instrument for better accuracy by moving the grating mount back and forth a little until the reading is as close to 5461Å as possible.

Measure the wavelength of all the lines in the spectrum you are observing. As usual, make several measurements of each line, average the results, and estimate the error of your measurements.

Kit Inquiry 13-2c What wavelength did you obtain for the blue line of mercury? For the green line? For the yellow line? For the red line (if it was observable)?

Kit Inquiry 13-2d What were the errors of each wavelength you determined? Are they consistent with the laboratory values of the wavelengths or do there appear to be additional systematic errors? (The Mercury wavelengths are 4358Å, 5461Å, and 5790Å.)

C. Observing Other Light Sources

See whether you can find other sources of light to measure. There are several types that would be quite instructive to observe.

Kit Inquiry 13-2e Observe a fluorescent lamp. (NOTE: You will need to darken the room to gain the contrast necessary to see the spectrum clearly, or perhaps stand in a different room from the lamp.) Can you see any evidence of a continuous spectrum? Can you see any evidence of discrete emission features? If so measure their wavelengths and try to identify the gas in the tube.

Kit Inquiry 13-2f Try to find a pinkish street lamp to observe at night. Note the wavelengths of any features you can see in its spectrum. Can you find an example of what is called a **dark line** in this spectrum; that is, a wavelength where energy is being absorbed rather than emitted? At what wavelength was the dark line? The gas in this tube is sodium.

Many advertising signs, particularly in the windows of business establishments, consist of a glass tube filled with a gas through which a high-voltage electric current flows. They are commonly called neon signs, although several different gases are used in addition to neon. Find several such signs of different colors and examine them with your spectrometer. You will find this easiest if you stand at one end of the sign and view it edgewise, so that the entire sign appears as a thin, vertical line, rather than standing in front of the sign as you would read it (**Kit Figure 13-2-5**). This will give you a thin source that will make a good, sharp spectrum after the light is dispersed by the grating, and it reduces the confusion that results from nearby lights.

Kit Inquiry 13-2g Can you find a red neon light to observe? If so, measure its wavelengths and compare them to your mercury vapor observations.

Kit Inquiry 13-2h If you have any colored pieces of plastic, it is instructive to observe an ordinary light bulb through them. Describe the effect that each has on the light passing through it by contrasting the appearance of the spectrum of the light bulb with and without the plastic in front of it.

D. Observing the Solar Spectrum

Because the Sun is a large ball of hot, dense gas, it gives off a continuous spectrum. It is also surrounded by a cooler, outer atmosphere that absorbs energy at certain wavelengths, giving a dark-line spectrum superimposed on the continuous spectrum.

You can safely observe the Sun with your spectrometer but to see the dark lines in the spectrum is a delicate observation, and you should follow the instructions carefully. Read them all the way through first.

This exercise should be done only in the late afternoon near sunset, when the Sun is just about to dip below the horizon. Find a location to observe where there is a relatively clear western horizon. At this time, sunlight passes through more atmosphere than at other times, so the Sun will be fainter than earlier. You will also find it easier to hold your spectrometer still in a horizontal position than if you had to point it up in the sky. Narrow the slit of the spectrometer until it is extremely narrow—a fraction of a millimeter. The narrower you can get it and still see the spectrum, the better it will work. When the spectrometer is pointed at the Sun, most of the visible disk of the Sun is

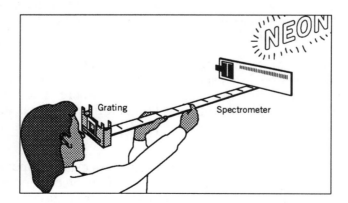

Kit Figure 13-2-5 Technique of observing a "neon" sign from the sign rather than directly in front.

blocked by the spectrometer scale, and only a thin rectangle of solar light is passed by the slit to reach your eye. When you look through the grating, you should have no difficulty in seeing the continuous spectrum of the Sun spread out along the scale.

Finally, to succeed in seeing the dark lines, you will have to rest the spectrometer on something so that it does not shake. Use a table, tape it to a tripod, or whatever, but do not hand-hold it.

With the solar continuous spectrum in view on the scale, narrow the slit slowly and carefully until you notice thin, *vertically* oriented dark lines in the spectrum. These are the absorptions by the cool gases of the solar atmosphere. For example, you should see a dark line in the yellow portion of the spectrum This feature is one of the most visible and is caused by sodium gas in the Sun.

Kit Inquiry 13-2i What are the wavelengths of the dark lines you observe?

Kit Inquiry 13-2j Which of these lines, if any, are due to hydrogen in the solar atmosphere?

The spectra of all stars exhibit dark lines of this type, and it is by identifying their wavelengths and measuring line strengths that astronomers can determine the chemical composition of the stars.

Kit Inquiry 13-2k How might you determine which dark lines in the spectrum are due too the Sun and which are due to absorption by the Earth's atmosphere?

Answers to Kit Inquiries

13-2a. About 40 Å.

13-2c. The wavelengths are: Blue = 4358Å, green = 5461Å, and yellow = 5780Å.

13-2f. The dark line is at 5890Å. Note how its appearance differs from those places in the spectrum that just don't have any light emitted (i.e., the spaces between the bright lines).

Part III
Discovering the Techniques of Astronomy

Activity 13-2   Analyzing Light Sources With a Spectrometer

Answer Sheet   NAME _____

Inquiry 13-2a:

Inquiry 13-2b: (You may find that drawing a picture works better than words.)

Inquiry 13-2c, d: Record your measurements of the emission lines in the Mercury spectrum here. You should observe each wavelength at least three independent times and average your values.

Wavelength	Mercury spectral lamp #1	#2	#3	#4	#5	Ave.	Error
_____	____	____	____	____	____	____	____
_____	____	____	____	____	____	____	____
_____	____	____	____	____	____	____	____

Inquiry 13-2e:
(continued on next page)

Wavelength	Fluorescent spectral lamp #1	#2	#3	#4	#5	Ave.	Error
_____	_____	_____	_____	_____	_____	_____	_____
_____	_____	_____	_____	_____	_____	_____	_____
_____	_____	_____	_____	_____	_____	_____	_____
_____	_____	_____	_____	_____	_____	_____	_____
_____	_____	_____	_____	_____	_____	_____	_____

Inquiry 13-2f:

Wavelength	Pinkish spectral lamp #1	#2	#3	#4	#5	Ave.	Error
_____	_____	_____	_____	_____	_____	_____	_____
_____	_____	_____	_____	_____	_____	_____	_____
_____	_____	_____	_____	_____	_____	_____	_____
_____	_____	_____	_____	_____	_____	_____	_____
_____	_____	_____	_____	_____	_____	_____	_____
_____	_____	_____	_____	_____	_____	_____	_____

Do you see any new types of features in this spectrum that have not been seen in the other spectra? If so, at what wavelengths?

Inquiry 13-2g:

Wavelength	Red Neon spectral lamp #1	#2	#3	#4	#5	Ave.	Error
_____	_____	_____	_____	_____	_____	_____	_____
_____	_____	_____	_____	_____	_____	_____	_____
_____	_____	_____	_____	_____	_____	_____	_____
_____	_____	_____	_____	_____	_____	_____	_____

Inquiry 13-2h:

Kit Activity 14-1

Classification Using
Known Spectra

When you have completed this activity, you should be able to do the following:

• Classify an unknown spectrum to within one-half a spectral class, given a sequence of known comparison spectra.

• Discuss the characteristics of spectra of stars having different temperatures.

Astronomers traditionally studied stellar spectra by placing photographic film at the focus of the spectrograph. Nowadays, however, an electronic detector replaces the photographic plate. In the activity, you will classify spectra obtained with these modern techniques.

You are to compare the unknown spectra in Kit Figure 14-1-1 with the standards in Kit Figure 14-1-2. None of the unknown spectra will be exactly like the comparison spectra, because they are for different stars; some will be somewhat similar and others will be quite different. Attempt to classify each spectrum by finding the standard spectrum that most closely resembles the unknown spectrum, feature for feature. Write down the spectral type of the spectrum that gives you the best match. If the unknown spectrum clearly lies about halfway between two known spectra, write down the spectral type that is about halfway between the two known spectra (e.g., if your unknown spectrum lies halfway between A0 and A7, then you could write down either A3 or A4).

You are looking for two types of characteristics. First, look at the general appearance of the spectrum. The overall shape is an approximation of the continuous spectrum emitted by the star. It is an approximation because the overlying atmosphere modifies the continuous spectrum in producing the absorption lines, which are the second type of characteristic you examine. You are looking for specific spectral features that characterize each spectral type. Are lines of neutral or ionized helium present when the hydrogen Balmer lines are weak or absent? Then the spectrum is an O or B type, if you identified the helium lines correctly. Are the hydrogen lines the dominant feature? Then the spectrum is probably of type A. How about metals? Molecular bands? Then the star is cooler, probably of type F, G, K, or M, depending on what is present and how strong it is. In this way you can make a rough classification.

To match the spectra more exactly, you will need to look at more subtle characteristics. For example, in stars of types A to G, you can compare the strengths of the hydrogen Balmer line (Hγ) at 4340 Å and the ionized calcium line at 3933 Å. The hotter the star, the weaker the calcium line will be in comparison to the hydrogen line. At spectral type A7, they are about equal in strength, whereas in F and G stars, the line of ionized calcium is stronger.

In a similar way, you can compare the strength of the neutral calcium line at 4226 Å and of the group of iron lines at 4271 Å with the hydrogen line at 4340 Å. The iron and calcium lines will become more and more prominent as the star gets cooler. Similarly, the titanium oxide (TiO) bands near 4761 and 4944 Å can be compared with the calcium and iron lines if the star is very cool, whereas in O and B stars, you can compare the strengths of the hydrogen, helium, and ionized helium lines.

Summarize your results. Include information about what features you found and the relative strengths of the various features that guided you to the classification you made.

Kit Inquiry 14-1a Summarize in your own words the characteristic features of each different spectral type, referring only to Kit Figure 14-1.

Kit Inquiry 14-1b What is the contradiction if a stellar spectrum exhibits both molecular lines and helium lines? How might such a contradiction be resolved?

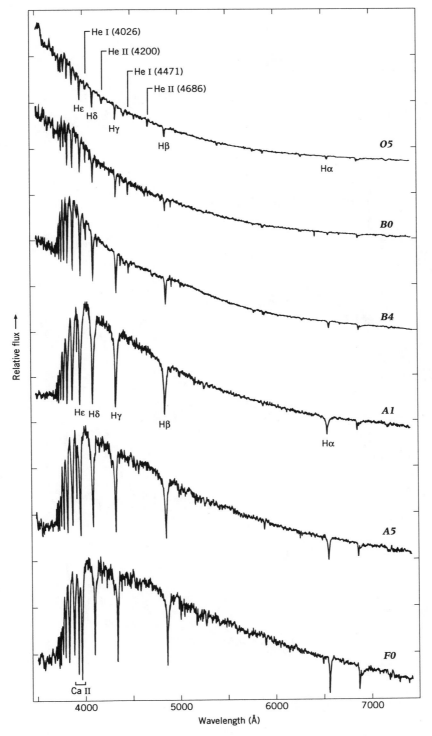

Kit Figure 14-1-1a Standard stars of known spectral type.

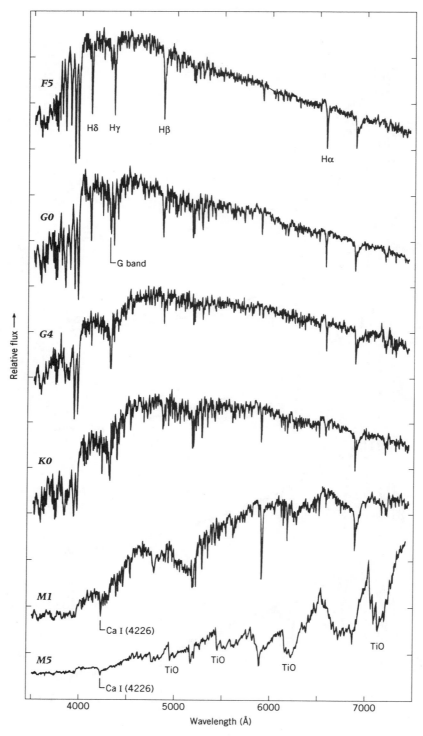

Kit Figure 14-1-1b Standard stars of known spectral type.

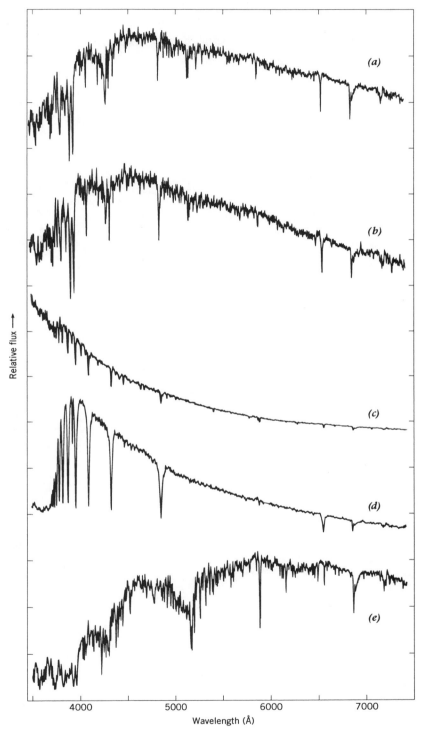

Kit Figure 14-1-2 Stars for spectral classification.

Part IV
Discovering the Nature and Evolution of Stars

(The spectra in this Activity are from *A Library of Stellar Spectra* by Jacoby, Hunter, and Christian. Courtesy of National Optical Astronomy Observatories.)

Kit Activity 19-1

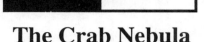

The Crab Nebula

When you have completed this activity, you should be able to do the following:

• Determine the scale of a photograph, given the angular separation of two objects.

• Determine the scale of a spectrum, given the wavelengths of two or more spectral lines.

• Determine the velocity of expansion of the Crab Nebula from a spectrogram.

• Determine the distance material from the supernova remnant has flowed since the time of the supernova explosion.

• Determine the distance to the Crab Nebula.

Use the answer sheet at the back of this activity, and show your work on the back of it.

You have read that the Crab Nebula is a remnant of a supernova explosion thought to have occurred in the year 1054. The expansion of the nebula can be observed using two techniques: From photographs taken some years apart, and from the spectrum.

Part I. The Photographs

Examine the two photographs in **Kit Figure 19-1-1**, which were taken in 1942 and 1976. Find some small, well defined knots or sharp-edged filaments of gas near the edge of the nebula that have clearly moved. If you cannot find any, ask your instructor.

• **Kit Inquiry 19-1a** Measuring to a tenth of a millimeter, how far has the knot you are considering moved between the times when the photographs were taken?

• **Kit Inquiry 19-1b** How many millimeters is the knot from the center of the nebula?

• **Kit Inquiry 19-1c** Use your results from the previous two questions to determine how many years it took the knot to get from the supernova explosion to its current location.

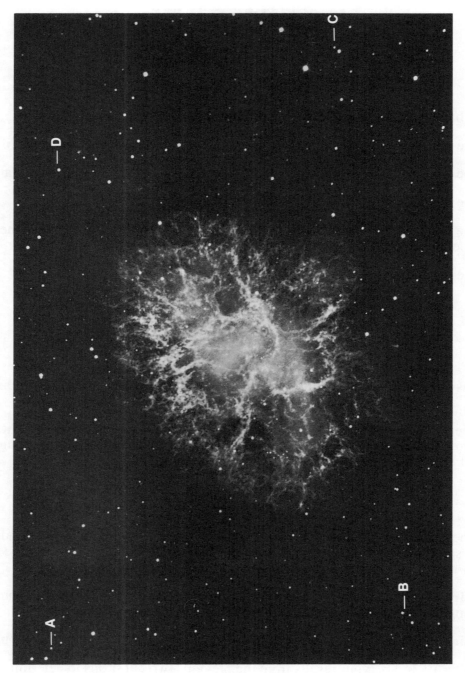

Kit Figure 19-1-1a The Crab Nebula as observed in 1942.

Kit Figure 19-1-1b The Crab Nebula as observed in 1976.

•**Kit Inquiry 19-1d** What assumption are you making?

• **Kit Inquiry 19-1e** According to your data, in what year did it occur?

• **Kit Inquiry 19-1f** Compute the percentage error in your answer from the year 1054.

Part II. The Spectrum

The spectrum, shown in **Kit Figure 19-1-2**, contains the spectrum of both the nebula (the curved lines) and the spectrum of mercury that comes from lamps in the nearby city. In this instance, we can take advantage of the mercury lines and use them in our activity. As shown by the mercury wavelengths, wavelength changes along the horizontal direction.

The nebular spectrum is an emission-line spectrum obtained at Lick Observatory near San Jose, California. The brightest line is a forbidden line of ionized oxygen [O III] at a wavelength of 3727Å. It has the appearance it does because light from gas coming toward the observer is blueshifted, while light coming from gas moving away from the observer is redshifted. The separation of the blueshifted and redshifted components of the 3727Å line is, therefore, related to the expansion velocity of the nebula.

• **Kit Inquiry 19-1g** Carefully measure the number of millimeters between the mercury lines of wavelength 3650Å and 4358Å.

• **Kit Inquiry 19-1h** How many angstroms are there in each millimeter on the photograph of the spectrum?

• **Kit Inquiry 19-1i** Measure, in millimeters, the maximum separation between the two components of the 3727Å emission line.

• **Kit Inquiry 19-1j** What is the separation of the two components in angstroms?

• **Kit Inquiry 19-1k** Use the Doppler shift formula to find the velocity of expansion of the nebula. The result of your calculation will be the velocity of one side relative to the other side. To find the expansion rate relative to the supernova in the middle, you need to divide the result of your calculation by two. Place your result on the answer sheet.

• **Kit Inquiry 19-1l** Using this expansion velocity, how far in kilometers has the knot you measured in Part I moved since the supernova occurred?

Part III. Distance

In Activity 3-1 you first encountered the formula relating angular size, linear size, and distance. You will now apply this formula to determine the distance to the Crab Nebula.

To determine the angular distance of your knot from the center, you must first determine the scale of the photograph you used in Part I. The scale tells you the number of seconds of arc covered by each millimeter on the photograph. You can do this by

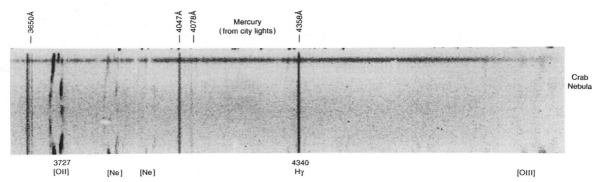

Kit Figure 19-1-2 A spectrum of the Crab Nebula. The straight lines are caused by
Mercury lamps in a nearby city. The curved lines are from the nebula.

measuring the separation in millimeters between stars of known separation, such as those
marked A-D on the photograph. The separations of the marked pairs, in seconds of arc
are as follows:

Star pair	Separation seconds of arc
AB	351
AC	689
AD	500
BC	596
BD	576
CD	302

- **Kit Inquiry 19-1m.** What is the scale of the photograph in seconds of arc per
  millimeter? Find it by averaging the scales determined from each of the measured
  separations.

- **Kit Inquiry 19-1n.** What is the distance in mm of your knot from the center of the
  nebula? How many seconds of arc is this? Express your result in degrees.

- **Kit Inquiry 19-1o.** Now that you know the angular distance through which the knot has
  moved, as well as the linear distance it has traveled (from Kit Inquiry 19-1l), determine
  the distance to the Crab Nebula in kilometers and light years.

- **Kit Inquiry 19-1p.** Use whatever methods you like to try to determine which star is the
  neutron star in the Crab Nebula.

Part IV
Discovering the Nature and Evolution of Stars

The photographs of the Crab Nebula were kindly supplied to us by Owen Gingerich and Sky Publishing Corporation. A more thorough exercise on the Crab Nebula is in *Laboratory Exercises in Astronomy -- The Crab Nebula*, published in *Sky and Telescope* November 1977, and available from Sky Publishing Corp.

Activity 19-1   The Crab Nebula

Answer Sheet   NAME _____

Show your calculations on the backside of the answer sheet.

Part I. The Photographs

Kit Inquiry 19-1a. Distance knot has moved: _____ mm

Kit Inquiry 19-1b. Distance of the knot from center of the nebula: _____ mm

Kit Inquiry 19-1c. Time since explosion: _____ years

Kit Inquiry 19-1d. Assumption made: _____

Kit Inquiry 19-1e. Year of explosion: _____

Kit Inquiry 19-1f. Percent error:_____

Part II. The Spectrum

Kit Inquiry 19-1g. Separation of mercury lines: _____ mm

Kit Inquiry 19-1h. Ångstroms per millimeter: _____

Kit Inquiry 19-1i. Separation of blue and red components: _____ mm

Kit Inquiry 19-1j. Separation of blue and red components: _____ Ångstroms

Kit Inquiry 19-1k. Expansion velocity relative to supernova: _____ km/sec

Kit Inquiry 19-1l. Distance know has moved: _____ km

Part III. Distance

Kit Inquiry 19-1m. Scale of photograph _____ seconds of arc/mm

Kit Inquiry 19-1n. Distance of knot from center: ___ mm; ___ seconds of arc;___ degrees

Kit Inquiry 19-1o. Distance _____ km;_____ ly

Kit Inquiry 19-1p. Drawing to show your estimate of which star is the neutron star (over).

Part IV
Discovering the Nature and Evolution of Stars

Kit Activity 21-1

Classification of Galaxy Types
from Sky Survey Photographs

When you have completed this activity, you should be able to do the following:

• Classify galaxies on the basis of their appearance on photographs.

• Estimate the proportions of the various galaxy types in Virgo and discuss the source of the uncertainties in these proportions.

Use the answer sheet at the back of this activity.

For this activity, use **Kit Figure 21-1-1**, a photograph of a region of the sky in the constellation of Virgo taken with the 1.2-meter Schmidt telescope at Mount Palomar Observatory. Schmidt telescopes allow both short exposure times and a large field of view, thus making them ideal instruments for comprehensive surveys of the sky. This particular telescope has mapped all the sky visible from Mount Palomar, from the North Celestial Pole down to 24° South of the celestial equator, on large glass photographic plates, each covering a 6° by 6° area of the sky. On the scale of the original photograph, 1 mm equals 67 seconds of arc. Centimeter marks and numbers (with half-centimeter ticks) are indicated along the edges of the figure for specifying any position on the photograph. For example, at position $X = 9.6$, $Y = 18.6$ (which means 9.6 cm from the left edge of the photograph and 18.6 cm from the bottom), there is a prominent spiral galaxy. You can examine the objects in the photograph in more detail by *using the lenses from your experiment kit as magnifiers.*

Distinguishing Stars From Galaxies

The photograph contains both bright and faint stars and galaxies. The brighter stars have "spikes" around their images; these are caused by light diffracted by the supports that hold the film in place. Galaxies, which are extended objects rather than point sources, do not show such spikes. However, many stars are too faint to have visible spikes, so you must distinguish faint stars from faint galaxies by the sharpness of the edge of the image. Examine a few faint images on the photograph and see whether you can make this distinction. The faintest images on this photograph are about 100 million times fainter than the bright stars Arcturus and Vega.

Setting Up a Classification Scheme

The first step in trying to understand nature is to set up a classification scheme that divides objects into categories. After some experience, we hope to be able to make sense out of the categories. The activity of classifying is important in itself because it sharpens our powers of observation and helps spot important variations between objects. For example, in Kit Figure 21-1-1 there is a definite division between the objects that show a spiral form and those that do not. Within these two categories, however, we can further distinguish among objects on the basis of their appearance. For example, we could compare the relative sizes of the nucleus of spiral galaxies and their spiral arm regions, or we could distinguish different shapes.

• **Kit Inquiry 21-1a** Examine the spiral galaxies in Kit Figure 21-1-1. In what ways do they differ from each other? List examples of different types of spirals, identifying them by their X and Y coordinates.

• **Kit Inquiry 21-1b** Can you find edge-on examples of each of the different classes of spirals?

• **Kit Inquiry 21-1c** How do the nonspiral galaxies differ from each other?

• **Kit Inquiry 21-1d** In Kit Figure 21-1-1, count all the galaxies in the region enclosed by dashes. Keep track of spirals and nonspirals separately, recording your results on the answer sheet. From these counts, what appears to be the relative proportion of spiral to nonspiral galaxies? (If you cannot tell whether a galaxy is spiral or elliptical, *do not use it* in your estimate of the ratio of spirals to ellipticals. You may, however, use it in your overall estimate of the number of galaxies.) How many galaxies of all types do you think might be visible in the entire photograph? How did you make your estimate?

Part V
Discovering the Nature and Evolution of Galaxies and the Universe

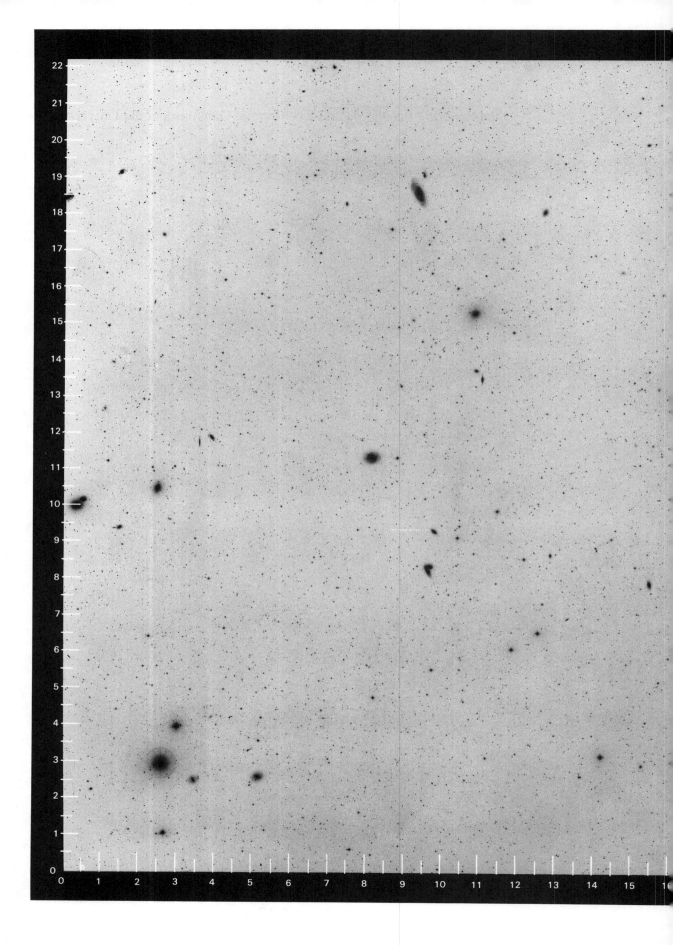

Kit Figure 21-1-1 A portion of a Sky Survey photograph showing a region in the Virgo cluster of galaxies.

Part V
Discovering the Nature and Evolution of Galaxies and the Universe

Activity 21-1   Classification of Galaxy Types from Sky Survey Photographs

Answer Sheet   NAME _____

Show your calculations on the backside of the answer sheet.

Inquiry 21-1a: Spiral galaxy examples

x-coordinate	y-coordinate	x-coordinate	y-coordinate
_____	_____	_____	_____
_____	_____	_____	_____
_____	_____	_____	_____
_____	_____	_____	_____
_____	_____	_____	_____

How they differ from each other:

Inquiry 21-1b: Edge-on examples

x-coordinate	y-coordinate	x-coordinate	y-coordinate
_____	_____	_____	_____
_____	_____	_____	_____

Inquiry 21-1c:

Inquiry 21-1d: Galaxy counts

Spirals: _____

Non-spirals: _____

Proportion of spirals/non-spirals: _____

Estimate of total number of galaxies: _____

Technique used to make estimate:

Kit Activity 21-2

The Variation of Galaxy Sizes

When you have completed this activity, you should be able to do the following:

• Determine the relative sizes of galaxies in a cluster.

• Estimate the distance of one galaxy cluster relative to that of another.

Use the answer sheet at the back of this activity.

Return to Kit Figure 21-1-1, the photograph of the Virgo region. Consider now the following additional fact: Most of the galaxies you see in that photograph are members of a large *cluster* of galaxies. This "Virgo Cluster" is approximately 40 million light-years away and contains at least a thousand galaxies; it is several million light-years in diameter. The cluster is small compared to its distance, so we can assume that all the galaxies in the cluster are approximately the same distance away from us.

In Kit Activity 21-1, it was not possible to tell if the small galaxies in the photo were intrinsically small or merely farther away. But with this new information---that all the objects are at approximately the same distance---we can now argue that their differences in apparent sizes are proportional to the differences in their actual sizes.

Examine the photograph and determine the range of sizes exhibited by the spirals and the ellipticals. Use your lenses as magnifiers to help. In millimeters (or fractions thereof), what are the sizes of the largest spirals and ellipticals on the photograph, and what are the sizes of the smallest spirals and ellipticals? (Do not force yourself to classify a galaxy if its image is so small that you cannot really tell what it is.) The variation in apparent sizes that you see here must be indicative of the true range of variation of intrinsic sizes.

• **Kit Inquiry 21-2a** If we were to assume that all spirals had the same size, and that all ellipticals had the same size, how much of an error in the distance to an individual galaxy might we make?

Part V
Discovering the Nature and Evolution of Galaxies and the Universe

• **Kit Inquiry 21-2b** A photograph of a cluster of galaxies in Hercules is shown in **Kit Figure 21-2-1**. Both clusters are photographed on the same scale. It is apparent that the galaxy images in the Hercules cluster are much smaller than those in Virgo. How much smaller do the galaxies appear in Virgo than in Hercules? If the Virgo cluster is 40 million light-years away, how far away would the Hercules cluster be? What assumption is implicit in this comparison?

Kit Figure 21-2-1 (*a*) A cluster of galaxies in Hercules, to the same scale as Kit Figure 21-1-1. (*b*) The central portion of the cluster enlarged by a factor of eight.

Part V
Discovering the Nature and Evolution of Galaxies and the Universe

---

Activity 21-2   The Variation of
                                          Galaxy Sizes

Answer Sheet   NAME _____

Show your calculations on the backside of the answer sheet.

1. Largest Spirals on photograph (mm)

   x-coordinate   y-coordinate   mm

   _____   _____   _____

   _____   _____   _____

   _____   _____   _____

   _____   _____   _____

   _____   _____   _____

2. Largest ellipticals on photograph (mm)

   x-coordinate   y-coordinate   mm

   _____   _____   _____

   _____   _____   _____

   _____   _____   _____

   _____   _____   _____

   _____   _____   _____

(continued on next page)

3. Smallest Spirals on photograph (mm)

x-coordinate	y-coordinate	mm
_____	_____	_____
_____	_____	_____
_____	_____	_____
_____	_____	_____
_____	_____	_____

4. Smallest ellipticals on photograph (mm)

x-coordinate	y-coordinate	mm
_____	_____	_____
_____	_____	_____
_____	_____	_____
_____	_____	_____
_____	_____	_____

Inquiry 22-1a:

Inquiry 21-1b:

Appendix

Activities and Observations with a Small Telescope

He got a good glass for six hundred dollars

His new job gave him leisure for star-gazing

Often he bid me come and have a look

Up the brass barrel, velvet black inside,

At a star quaking in the other end.

Robert Frost, "Star Splitter," 1923

This appendix contains a sampling of projects that can be carried out with a small telescope or binoculars. Most can be done with the telescope that you made with your kit, although a few need a somewhat larger instrument (2.4 inches or larger in aperture). All the appendix activities are dependent on good weather and some will require observations that are planned over a period of time.

Each activity in the appendix is independent and has its own objectives. You may do as many as you have the time, inclination, or equipment to do. The following section, however, presents some general considerations on observing techniques and should be read first, since it applies equally to all the activities. In addition, if you have access to a larger telescope, you should by all means use it, because it will allow you to make observations that are better in every way.

This appendix includes activities on Mapping the Moon, Observing the Sun, Determining the Diameter of the Sun, Observations of Stars, Clusters, and Nebulae, and Observing the Planets.

General Considerations

The following general observing tips are particularly useful for naked-eye observing, but are also relevant when using a telescope.

1. *Observe under a dark sky.* City lights drown out the fainter objects.

2. *Adapt to the dark.* Relax for a few minutes until your pupils dilate. Then you will be able to see more.

3. ***Use a red flashlight.*** A red flashlight allows your eyes to remain dark adapted whereas a white light will cause your pupils to contract.

4. ***Use averted vision.*** Your eye has a region of higher sensitivity away from the middle of the retina. Therefore, looking slightly to the side is often more effective in seeing faint objects than looking directly at it.

5. ***Take your time.*** If you rush, you will miss observing many interesting things and not enjoy your time under the stars as much.

6. ***Move from the known to the unknown.*** Start with constellations and stars you recognize or can readily find on your charts. You will then be able to move step by step to new objects.

7. ***Make reliable records.*** Write down everything; do not leave anything to memory. In making good records, you will take your time and enjoy yourself much more.

Additional suggestions that apply particularly to telescopic observations follow.

1. ***Keep Your Telescope as Steady as Possible.*** Always rest the telescope against something sturdy--a wall, a tree, or something similar. In this way you will reduce hand quivering, which of course is magnified by the telescope. Better yet, make a crude mount for any small telescope by placing it between two *heavy* books and wrapping stout rubber bands around them, as shown in **Kit Figure A-1**. If the assembly is placed on a solid table or ledge, you can look through the telescope without touching it, once it is pointing in the right direction, with greatly improved steadiness. (However, this arrangement will not work too well near the zenith.) For best results use a tripod.

2. ***Image Reversal.*** Most telescopes designed for celestial (as opposed to terrestrial) viewing invert the image and also reverse it left to right. Binoculars do not have this problem. If you are using such a telescope, it can be confusing at first. One easy way to get used to this is to practice locating the Moon with the telescope, keeping *both* eyes open to compare the direction you move the telescope with the direction that stars appear to move across the field of view. When this can be done easily, practice with bright stars and planets; you will quickly be able to locate even faint stars with ease.

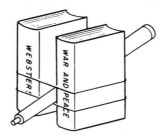

Kit Figure A-1 How to steady a small telescope with two
heavy books bound with string or strong rubber bands.

Appendix
Activities and Observations with a Small Telescope

You can judge directions in the telescope by moving it in a certain direction--for example, toward the east--and seeing toward which direction the stars appear to move. You should always include the directions on the plane of the sky whenever you draw a sketch of any object.

• **Kit Inquiry A-a** Face south with the telescope or binoculars in hand. East is to your left. Viewed in the telescope, which way is east? Which way is south?

3. ***Resolving Power and Seeing.*** The ability of a telescope to reveal fine details in an image--its resolving power--depends on two things: Its aperture and the steadiness of the atmosphere when the observations are made. Theoretically, a 2.5-cm (1-inch) telescope should be able to distinguish details that are about 4 seconds of arc apart, whereas the resolving power of a 10-cm (4-inch) telescope, a common size for a small telescope, should be about 1 second of arc. However, the perfect viewing conditions required for a telescope's theoretical resolving power to be attained are seldom encountered, even with an optically perfect and rigidly mounted telescope.

When the atmosphere is turbulent, it is said that one has a night of bad "seeing." In practice, you will probably notice this first while attempting to observe close binary stars. Under these conditions, the images of the two stars become badly blurred and will blend together, appearing to be one star.

4. ***Light-Gathering Power.*** The aperture of a telescope is important for another reason. Since it gathers more light than the dark-adapted eye, one can see objects that are many times fainter than those that can be observed with the naked eye. Under perfect conditions--dark, clear skies, far from city lights--stars of magnitude 9.0 could be seen with a small opera glass. With a 10-cm telescope objects with magnitude 12.0 would be observable. However, perfect conditions are rarely encountered. In general, you should be able to see objects about three magnitudes fainter through a 2.5-cm telescope than with the naked eye, so that on a night when fourth-magnitude stars are visible, you should be able to see down to magnitude 7 with a small pair of binoculars.

5. ***Magnification.*** Do not use excessively high magnifications! If your telescope comes with various eyepieces, so that you can use it at several different magnifications, you will find the *lowest* magnification to be the most generally useful one. At lower magnifications you will find the images steadier and brighter, though smaller than at higher magnifications. Use higher magnifications on an object only after you have fully exploited the lowest magnification eyepiece, and even then proceed in stages. You will find that the highest magnifications are useless except under the very best conditions of atmosphere and dark sky.

6. ***Field of View.*** It will be handy for you to determine the field of view of your telescope so that you can estimate the angular sizes of objects by noting the fraction of the field of view that they occupy in the telescope. There are several ways to do this.

a. ***Outside Method*** View the Moon when it is at least in its quarter phase. Focus carefully and then estimate how many images of the Moon will fit in the field of view, using the *longest* dimension of the Moon. The Moon itself subtends half (0.5) a degree, so the field of view can be determined in a straightforward way.

For example, if four images of the Moon would just stretch across the field of view, then it would be $0.5° \times 4 = 2°$.

b. ***Indoor Method (can be done in bad weather)*** This book's longest dimension is 11 inches. Prop it against the wall and back away from it until the longest dimension just stretches across the field of view. Be sure to focus the telescope carefully--it will change slightly as you back away from the book. Measure your distance from the book in feet. Now use the angular size formula

$$\text{Angular size of field of view} = 57.3° \times \frac{\text{longest dimension of book}}{\text{distance from book}}.$$

• **Kit Inquiry A-b** What is the field of view of the telescope in degrees?

• **Kit Inquiry A-c** If you used more than one method to measure the field of view of the telescope, how well did they agree?

Special Note: If you are interested in astronomical observations more advanced than the ones in this book, consult the end of the appendix, where a number of references for observational astronomy are given.

Kit Activity A-1

Map of the Moon

When you have completed this activity, you should be able to do the following:

• Make a careful drawing of an astronomical object using the telescope.

• Identify major features on your drawing by comparison with a photograph.

Ever since the time of Galileo, one of the first things people have looked at when they obtained a telescope is the Moon. Galileo's drawings of the Moon (**Kit Figure A-1-1**) profoundly influenced his contemporaries and helped to usher in the new astronomy.

For this activity, you can use a pair of binoculars or a small telescope; you can use the telescope you constructed in the kit if nothing else is available. You will need to make observations at as many different phases of the Moon as possible, so it will take you several weeks to get everything together. At the very least, you should obtain observations at or near the crescent, quarter, gibbous, and full stages.

Make drawings of the Moon's face as accurately and detailed as you can. It will help if you try to draw only a small portion of the Moon at a time, and frequently check back and forth between your telescopic view and the picture you have drawn. Don't forget to write down the time and date of each observation. Brace the telescope as rigidly as you can; if a tripod is available, use it.

When you have completed a set of as many drawings as you can, combine them into a single *composite* drawing. It will help at this stage if you start all the drawings in your sequence with a circle that is 2 inches in diameter. Refer to your drawings and answer the following questions.

Indicate in your report the characteristics of the telescope you used for these observations. In particular, give its field of view and magnification.

• **Inquiry A-1a** Can more detail be seen on some parts of the Moon than on others? If so, which?

- **Inquiry A-1b** Can more detail be seen at some phases of the Moon than at others? Which ones? How can you explain this?

- **Inquiry A-1c** How many different kinds of features did you observe and map? Describe each of the different features you drew.

- **Inquiry A-1d** Another drawing of the Moon made by Christopher Scheiner at about the same time is shown in **Kit Figure A-1-2**. How do your drawings compare with Galileo's and Scheiner's? Compare your composite drawing of the Moon with a photograph of the Moon. How many of the named features on the Moon did you record? Which of the different types of features shown did you recognize?

- **Inquiry A-1e** If you used a 2-inch circle to draw the Moon, then 1 inch = 1000 miles in your drawings. Estimate the sizes of some of the features you drew. If you did not start with a 2-inch circle, adjust your calculation appropriately.

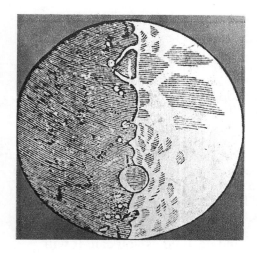

Figure A-1-1 An early drawing of the Moon by Galileo.

Figure A-1-2 An early view of the Moon drawn by Christopher.

Kit Activity A-2

Telescopic Observations of the Sun

When you have completed this activity, you should be able to do the following:

• Observe the limb darkening of the Sun and explain its cause and interpretation.

• Determine the Sun's rate of spin from observations of sunspot motions.

Warning: **Never look at the Sun through a telescope. Irreparable eye damage will result.**

Galileo observed the Sun directly through a telescope, and it has been suggested that this was a contributing factor to the blindness he suffered later in life. However, there is a safe way to project the image of the Sun onto a screen for observation that involves no risk to your vision.

Kit Figure A-2-1 illustrates the principle of eyepiece projection by which the Sun's image may be studied safely. The telescope is pointed toward the Sun (without looking through it!) and the Sun's image is projected back to a screen held some distance behind the telescope. The greater the distance between the telescope and the screen, the larger will be the Sun's image. The drawtube of the telescope must be moved in or out until the image is in sharp focus.

Although it is possible to hand-hold the telescope, a better method is to mount the telescope on a tripod (if available) or a support made of books (see Kit Figure A-1-1) to keep it steadily pointed at the Sun. It is also a good idea to cut a hole the size of the telescope tube in a good-sized piece of cardboard and slide this over the telescope to shield the screen from the direct rays of the Sun and increase the visibility of the image, as shown in **Kit Figure A-2-1**.

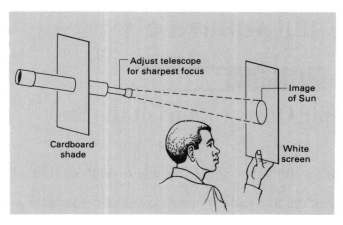

Figure A-2-1 Projecting the Sun's image onto a

To get the telescope pointed exactly toward the Sun, move it around (with the cardboard shield removed) until the shadow cast by the barrel of the telescope on the screen is *as small as possible;* you should then see the image of the Sun on the screen, although it may be out of focus. Put on the cardboard shield and then adjust the focus until the image is as sharp as possible. It will help if a friend assists you in this; one of you can adjust the telescope while the other holds the screen several feet away.

During the course of your observations you will have to change the direction in which the telescope is pointing from time to time to compensate for the daily motion of the Sun; however, you will not have to refocus the telescope.

• **Inquiry A-2a** Compare the brightness of the central part of the Sun's disk with that of the edge (or *limb*) of the Sun. Which one is brighter?

You should be able to discern that the edge of the Sun's disk is somewhat darker than the center, a phenomenon known as *limb darkening.*

This phenomenon enables us to make some interesting conclusions about the structure of the Sun's atmosphere. When we look at the edge of the Sun, we are looking at regions fairly high in the atmosphere, since the light from deeper layers is absorbed by the Sun's atmosphere before it can get out. But when we look at the center of the Sun's disk, light traveling the same distance through the atmosphere as at the edge comes from a deeper layer. Since the center is brighter, we can conclude that the region from which the light comes is hotter, on the average, than it is at the edge.

• **Inquiry A-2b** This simple observation of the differences in brightness of different parts of the solar disk shows us that the temperature of the Sun's atmosphere varies with depth. From the evidence of limb darkening, does the temperature *increase or decrease* with increasing depth?

By observing the Sun at various points around its surface, one is actually probing into different depths beneath the Sun. Of course, only the Sun is near enough for this to be done in detail, but the general result of increasing temperature with increasing depth applies to all stars. Limb darkening has been observed in a few stars--for example, with eclipsing binary stars, in which one star is gradually covered up by another--but this is the exception rather than the rule.

• **Inquiry A-2c** Can you see any sunspots on the surface of the Sun? Can you tell if the spots tend to be isolated or come in groups?

If you can detect sunspots, you will note that the same spots persist on the face of the Sun for a while. By observing the Sun each day you can determine its rate of rotation from the motions of the spots.

• **Inquiry A-2d** If you determined a rate of spin for the Sun, what did you obtain? Explain how you did the determination from your data.

Appendix
Activities and Observations with a Small Telescope

Kit Activity A-3

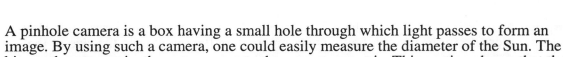

The Diameter of the Sun

When you have completed this activity, you should be able to do the following:

• Measure the diameter of the Sun using the pinhole experiment and explain how it works.

• Assess the effects that the pinhole shape and size have on the accuracy of the experiment.

A pinhole camera is a box having a small hole through which light passes to form an image. By using such a camera, one could easily measure the diameter of the Sun. The bigger the camera is, the more accurate the measurement is. This section shows that the same thing can be accomplished in a very simple way--using just a piece of paper with a hole in it, a small mirror, and a cardboard box.

Kit Figure A-3-1 is a pattern that shows the outlines of two "pinholes" that may be used to carry out the observations of this activity. They are in fact considerably larger than real pinholes, since one is a half-inch square and the other is a quarter-inch circle. We are going to construct a large pinhole camera by throwing our image over a large distance, so that the holes will still be small compared to the size of the camera.

Trace Kit Figure A-3-1 onto a sheet of paper. Use a sharp razor blade or a modeling knife (a pair of scissors will probably be too awkward here, unless you have some very tiny scissors) and *cleanly* cut out the two holes. Cut the paper containing the holes to the shape of your small mirror and tape it over the mirror to create two small mirrors of different shapes.

As a screen for viewing the image formed by the holes, use an ordinary sheet of white paper. Tape this viewing screen to the back of an ordinary cardboard carton (with the front end removed), as shown in **Kit Figure A-3-2**, and put the box on the ground so that the open face of the box points away from the Sun. The reason for placing the viewing screen inside a box is that the shading provided by the walls of the box increases the contrast of the image and makes it much easier to see.

Appendix
Activities and Observations with a Small Telescope

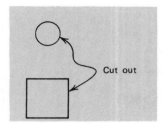

Kit Figure A-3-1 Pattern for the "pinholes" used in front of a mirror to project the Sun's image onto a wall.

Have a friend stand 3 or 4 meters away from the screen and hold the mirror so that the light from the Sun passes through the pinholes, is reflected off the mirror, and projected onto the screen (**Kit Figure A-3-3).** Cover first one hole, then the other. You should clearly see the image of the Sun projected by each of the "pinholes."

• **Inquiry A-3a** Does the square hole cast a square image and the round hole a round image, or does the shape of the hole seem to have no major effect on the shape of the image?

• **Inquiry A-3b** Which hole casts the *larger* image, the small one or the large one? How does the increase in the size of the image compare with the increase in the size of the hole?

• **Inquiry A-3c** Which image is brighter? To what do you attribute this?

• **Inquiry A-3d** Which image do you think would be easiest to measure an *accurate* diameter? Explain your reasoning.

Now measure the diameter of the sharper of the two images. Then measure the distance from the *pinhole* to the *screen.* Your friend can help you do this. Be sure both distances are expressed in the same units.

Kit Figure A-3-3 shows the geometry of the situation. The laws of similar triangles tell us that

$$\frac{\text{Diameter of Sun}}{\text{distance of Sun}} = \frac{\text{diameter of Sun's image}}{\text{distance of screen from pinhole}} .$$

If the Sun is 150 million km from the Earth, you can use the above ratio to compute the diameter of the Sun.

• **Inquiry 12-3e** What do you compute for the diameter of the Sun? How accurate is this determination?

Appendix
Activities and Observations with a Small Telescope

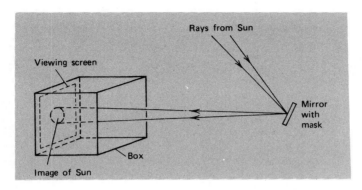

Kit Figure A-3-2 How to set up the solar diameter activity.

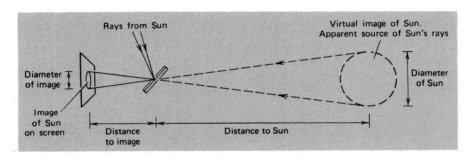

Kit Figure A-3-3 The geometry of the experiment. It is as if the Sun were behind the pinholes at the position of its virtual image. (Not to scale.)

Appendix
Activities and Observations with a Small Telescope

Kit Activity A-4

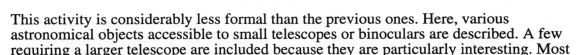

Observations of Stars, Clusters, and Nebulae

When you have completed this activity, you should be able to do the following:

• Describe a number of astronomical objects that you have observed through a telescope.

• Specify the time of year certain objects are visible in the sky.

This activity is considerably less formal than the previous ones. Here, various astronomical objects accessible to small telescopes or binoculars are described. A few requiring a larger telescope are included because they are particularly interesting. Most are readily within reach of even a 2.5-cm (1-inch) telescope.

The objects described include multiple stars, star clusters, bright diffuse nebulae, and external galaxies. The physical nature of each of these types of objects is described in greater detail later in the book, but a short description at this point would probably be useful.

DOUBLE AND MULTIPLE STARS, AND STAR CLUSTERS

Although to the naked eye it appears that most of the stars in the galaxy are single, it turns out that single stars are in the minority. Several stars that can be seen as double or multiple in a small telescope are pointed out and described in this section.

Star clusters are groups of stars that have a common origin, having been formed together by the collapse and fragmentation of a massive cloud of gas. They also share a common motion through the galaxy and are extraordinarily useful in studying the evolution of stars. Two main types of star clusters are distinguishable: galactic clusters and globular clusters.

Galactic clusters are relatively small and consist of only a few thousand stars. They are found generally near the plane of the galaxy (i.e., the Milky Way) and tend to be relatively young objects. On the other hand, globular clusters typically contain *several hundred thousand* stars, are often well out of the plane of the galaxy, and are among the oldest objects in the galaxy.

NEBULAE

Telescopes of respectable size were built long before astronomers learned how to distinguish among the different types of diffuse, nonstellar objects that are seen. All these were originally given the general description of "nebula," although it is now known that objects of dramatically different nature are included in this category.

Gaseous and diffuse nebulae are part of our own galaxy. They are large clouds of dust and gas whose presence is made known by nearby stars. An excellent example is the great Orion nebula, described among the January - February objects later in this section.

The category also includes the external galaxies, giant systems of stars like our own Milky Way galaxy. One of these is the great Andromeda galaxy; although it is composed of perhaps *one hundred billion* stars, we see it as only a small, fuzzy, and glowing patch of light in a small telescope because of its great distance of roughly 2 million light-years. A telescope of 100 inches in diameter or greater is required to see individual stars in such a distant system. The nearest of the external galaxies are the Large and Small Magellanic Clouds, which are visible only from the southern latitudes.

ARRANGEMENT OF THE DESCRIPTIONS

Objects will be described for the months when they are closest to the meridian, at around 8 P.M. (9 P.M. Daylight Saving Time). They can nevertheless be seen at other times of night and in other months. For example, the Pleiades Cluster is very close to the meridian in mid-January at 8 P.M. and is described among the January - February objects; it can be observed for several more hours as it moves toward the western horizon and could have been observed on the meridian at 10 P.M. in mid-December, 12 P.M. in mid-November, or 2 A.M. in mid-October. The star maps can be used to determine when objects are visible, or a celestial globe, set to the proper time and latitude, could be used instead.

Note that objects that are rather far south of the celestial equator (for observers in the Northern Hemisphere) will never rise very far above the horizon and should be observed as close to the meridian as possible, so that their altitudes will be at a maximum.

If you become a serious observer of the sky, you will rapidly exhaust the objects pointed out in this appendix. Your next step would then be to consult one of the many fine astronomical atlases available for observers. The most commonly used observing aid for amateur astronomers is Norton's Atlas.

DESCRIPTIONS OF SKY OBJECTS

Observations of the North Circumpolar Region

The star Mizar in the handle of the Big Dipper (**Kit Figure A-4-1**) is easily visible during most of the year and appears to be double when viewed with the naked eye. Its companion, Alcor, is not associated physically with Mizar and is actually very far from it in space; the two stars appear close to each other in the sky only by coincidence.

Binoculars will split the two easily.

It turns out that Mizar also has a true physical companion, about two magnitudes fainter than Mizar itself and about 14.5 seconds of arc distant from it. The two stars move in orbits about their common center of gravity and are the first true *physical* binary stars discovered. A small telescope will resolve this very nice pair.

January - February Objects
(Perseus, Auriga, Taurus, Orion)

h and χ Persei. This famous double cluster consists of a pair of galactic clusters just visible to the naked eye. In a small telescope they are a handsome sight. Like all clusters, the stars move around the galaxy together because they were formed from the same cloud of gas by a process of gravitational collapse and condensation. **Kit Figure A-4-2** shows the location of h and χ Persei in the sky.

The Pleiades. The most easily observed and widely known of the galactic clusters is the Pleiades, or "seven sisters," which are supposed to be the seven daughters of Atlas in mythology. If observed under the best conditions, eight or nine stars are visible to the naked eye; a small telescope reveals many more. This object is located in the shoulder of Taurus, the Bull, as illustrated in **Kit Figure A-4-3**.

M34. Another loose galactic cluster large enough for easy observation with a low-powered telescope is M34 (Kit Figure A-4-2). The designation M34 means that this object is the thirty-fourth in a list of "nebulous" objects compiled by Charles Messier, a French astronomer of the eighteenth century. Messier was searching for comets and in the process found many faint, fuzzy-looking objects. So that other comet watchers would not be distracted from their work by these objects, he published a list, intended to be a list of objects to *avoid* looking at. Nowadays, some of the most interesting objects in astronomy come from this list.

M36, M37, and M38. These are three galactic clusters found just south of α Aurigae (Capella), as shown in Kit Figure A-4-3.

M42 (the Great nebula in Orion). The central star in the sword of Orion (**Kit Figure A-4-4**) appears diffused to the eye, because it is not a star at all. Instead, it is an enormous region of gas and dust about 1500 light-years from the Earth, a region where star formation is believed to be going on at this very moment. It acquires its energy from the ultraviolet energy emitted by a cluster of bright stars that has formed out of the gas. Plate 11 is a photograph of this object taken with a large telescope, which reveals much more detail than can be seen in a small telescope.

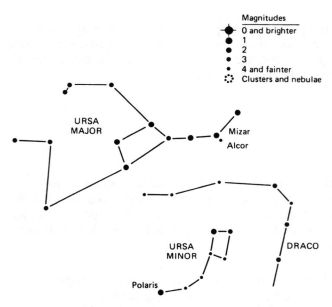

Kit Figure A-4-1 Alcor and Mizar in the Big Dipper.

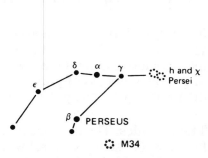

Kit Figure A-4-2 h and χ Persei and M34.

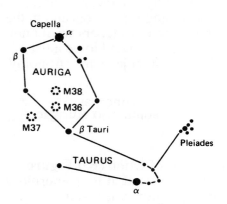

Kit Figure A-4-3 Taurus and Auriga. The Pleiades and other objects are also shown.

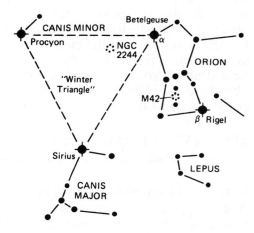

Kit Figure A-4-4 The Orion region: The Winter Triangle.

March - April Objects
(Gemini, Canis Major, Canis Minor, Cancer)

α **Geminorum (Castor).** This bright star is a double star that reached a separation of 6.5 seconds of arc in about 1880. The two components are now under 4 seconds of arc apart, so a telescope with several inches of aperture is needed to resolve the components into two separate objects. Spectral information shows that each of the components is itself a binary star, and there is yet a third, fainter component that is also a binary. Castor, therefore, actually consists of *six* separate stars. **Kit Figure A-4-5** shows how to find this object.

M35. This is a very beautiful galactic cluster, also located in Gemini (**Kit Figure A-4-5**).

M41. This galactic cluster can be found near Sirius (α Canis Majoris), as shown in **Kit Figure A-4-6.** Sirius, the brightest star in the sky, is itself a binary star, having a very faint white dwarf companion. A large telescope is needed to resolve this companion.

Kit Figure A-4-5 Gemini Kit Figure A-4-6 Canis Major.

NGC 2244. This galactic cluster is located between Orion and Canis Minor, as shown in **Kit Figure A-4-4.** The letters NGC refer to the New General Catalog, a list of nonstellar objects similar to Messier's list but much larger.

M44. One of the favorite galactic clusters in the sky, M44 is popularly known as Praesepe (the Beehive). **Kit Figure A-4-7** indicates where in the constellation of Cancer this object can be found.

ι **Cancri.** This is a binary system with contrasting yellow and blue stars at 4.4 and 6.5 magnitudes, separated by about 30 seconds of arc.

Appendix
Activities and Observations with a Small Telescope

May - June Objects
(Leo, Coma Berenices, Virgo)

γ **Leonis.** This is a famous double, at magnitudes 2.4 and 3.8, but separated by only 4 seconds of arc, so that a telescope with several inches of aperture is required to resolve it (**Kit Figure A-4-8**).

γ **Virginis.** This is a binary system with just under 6 seconds of arc separation, again requiring a moderate-sized telescope for resolution. What is unusual about this system is the fact that the brightness of the two components are very close, one at 3.6 magnitude and the other at 3.7. The two stars take about 180 years to make a complete orbit about each other; they were too close to be resolved in 1836 and will be so again in 2016 (**Kit Figure A-4-9**).

M3. This is a globular cluster, an enormous spherical system consisting of at least 100,000 stars. A small telescope will reveal a diffuse form and, in a telescope of 4 inches aperture or larger, a few individual stars can be resolved. Kit Figure A-4-9 shows where this object is located.

Coma Berenices. The stars of "Berenice's Hair" are actually an extended galactic cluster that is very beautiful when seen through field glasses or a small telescope. This region of the sky is also heavily populated with galaxies (although a small telescope will not give a satisfactory view of these distant systems), and the North Galactic Pole is in this region of the sky (**Kit Figure A-4-9**).

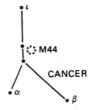

Kit Figure A-4-7 Cancer
and Praesepe.

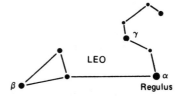

Kit Figure A-4-8 Leo.

July - August Objects
(Scorpius, Hercules)

M13. This object is located in Hercules (**Kit Figure A-4-10**) and is perhaps the most famous of the globular clusters. It appears as a diffuse patch in a pair of binoculars and will begin to be resolved in a telescope that has a few inches aperture.

Scorpius. This region is near the center of our galaxy and is rich in interesting objects, as shown in **Kit Figure A-4-11**. Look for *M80,* a bright globular cluster, and *M7,* a galactic cluster that is visible to the naked eye. For binary stars, depending on the observing conditions and your telescope, you can look for β Scorpii (magnitudes 2.9 and 5.2, separation 13.8 seconds of arc) or ν Scorpii (magnitudes 4.2 and 6.5, separation 41.5 seconds of arc). *Antares,* the very red star that is the brightest in the constellation, is a *supergiant* star so enormous that if it were placed where the Sun is it would extend all the way out to the orbit of Mars, engulfing the Earth. Antares is a binary star, but its companion is faint and close, and not resolvable in small telescopes.

September - October Objects
(Sagittarius, Lyra, Cygnus, Aquila)

Sagittarius. The center of the Milky Way galaxy is in this richly endowed region. Seek out the two gaseous nebulae shown in **Kit Figure A-4-11**: *M8,* the Lagoon nebula, a region of gas, dust, and star formation similar to the Orion nebula and faintly visible to the naked eye under good conditions, and *M17,* the Omega or Horseshoe nebula, which somewhat resembles the number "2" in shape. *M22* is the brightest globular cluster visible to Northern Hemisphere observers.

The Summer Triangle Region. At this time of year three bright stars--Vega (in Lyra), Deneb (in Cygnus), and Altair (in Aquila)--dominate the region near the zenith. The three stars form a conspicuous triangle known as the "Summer Triangle," which is a good starting point for finding other stars in the region (**Kit Figure A-4-12**). Several double stars are worth viewing in this region:

ε Lyrae. Two fifth-magnitude stars, separated by 208 seconds of arc, are easily seen in a small telescope. Larger telescopes will reveal that each of these is itself a binary system, making this object a "double-double" star.

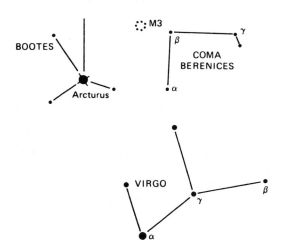

Kit Figure A-4-9 Virgo and Coma Berenices

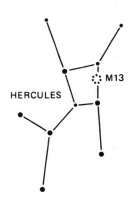

Kit Figure A-4-10 Hercules.

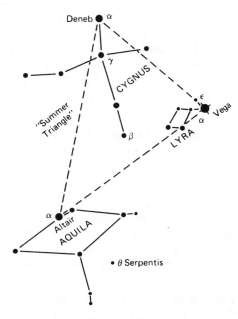

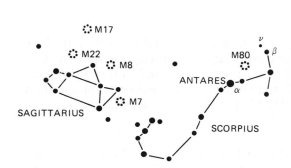

Kit Figure A-4-11 Scorpius and Sagittarius.

Kit Figure A-4-12 Lyra, Cygnus, and Altair: the Summer Triangle.

β **Cygni.** Separated by 35 seconds of arc, and with magnitudes of 3.0 and 5.3, the two stars in this system are of yellow and blue colors and contrast very nicely.

θ **Serpentis.** Near Aquila, this binary consists of two stars of magnitudes 4.0 and 4.2, separated by 22 seconds of arc.

November - December Objects
(Andromeda, Pegasus, Aries)

M31, The Great nebula in Andromeda. At a distance of 2 million light-years, this is the most distant object visible to the naked eye--an elongated oval patch of light located in Andromeda, as shown in **Kit Figure A-4-13**. It is actually a giant spiral galaxy, probably quite similar in structure to our own Milky Way galaxy, although photographs from large telescopes are required to reveal this fact. On a dark, clear night, you should be able to discern a diffuse, oval shape, sharply brighter in the central nucleus. This galaxy is one of the most important of the several dozen nearby galaxies, including our own, that are known collectively as the *Local Group*.

Two attractive binary systems within the region are (**Kit Figure A-4-13**):

γ **Andromedae.** Colored gold and blue at magnitudes 3.0 and 5.0, the stars are separated by 9.7 seconds of arc.

γ **Arietis.** Also a relatively close pair at a separation of 8.4 seconds of arc, the stars are of magnitudes 4.2 and 4.4.

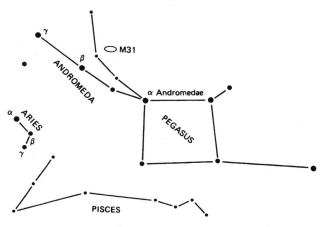

Kit Figure A-4-13 Pegasus, Andromeda and Aries.

South Circumpolar Objects

Kit Figure A-4-14 is a finding chart for objects within 30° of the South Celestial Pole. Of notable interest to observers in southern latitudes are:

47 Tucanae and ω Centauri. These are two magnificent globular clusters both visible to the naked eye. The cluster ω Centauri is a full 30 minutes of arc in apparent size; this is the apparent size of the visible disk of the Moon or Sun.

The Magellanic Clouds. These are two small galaxies that are so close to our own that they have significant gravitational interaction with it. They are only about 200,000 light-years distant from the Sun and are part 4 of the Local Group, along with M31. (Marked as SMC and LMC in Kit Figure A-4-14).

β **Tucanae.** This binary star has a separation of 27 seconds of arc and equally bright components of magnitude 4.5.

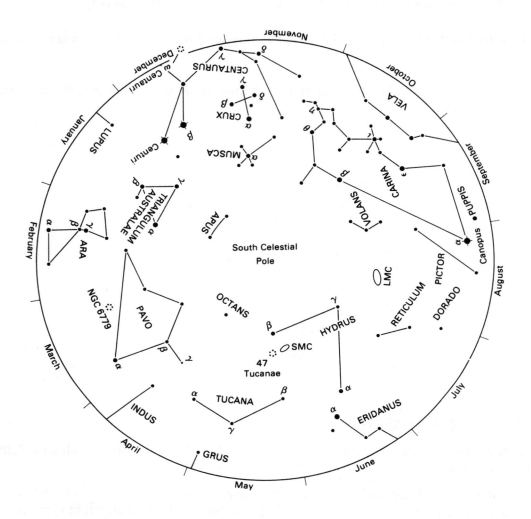

Kit Figure A-4-14 The south circumpolar region.

Kit Activity A-5

▮▮▮▯

Observations of the Planets

When you have completed this activity, you should be able to do the following:

• Discuss observations of the planets that you have made.

▮▮▮▯

Refer back to Activity Table 4-7-1 to locate the planet of interest. Observations that could be carried out over a period of time by a small telescope include the following.

• Observations of the motions of the Galilean satellites of Jupiter. They orbit Jupiter with periods ranging from hours to days. Using Newton's form of Kepler's third law, you can calculate the mass of Jupiter (or the mass of any planet for which you can see and measure the motion of a satellite).

• Observations of the surface banding on Jupiter and the rotation of the Great Red Spot. Jupiter's rapid rotation makes it noticeably flattened.

• Observations of the phase changes of Venus. It would take occasional observations spread out over several months to observe the variations in the phase of Venus that enabled astronomers to finally reject the Ptolemaic theory of the geocentric universe.

• Observations of Saturn's ring system. Observations over a period of years will show the changes in the inclinations of the rings. The rings are so thin that when they are edge-on to the Earth they disappear.

• Accurate observations of the motions of the planets with respect to the background stars, including periods of retrograde motion and determination of their rates of motion.

Answers to Kit Appendix Inquiries

A-a. East is on the right in an astronomical (inverting) telescope and on the left in binoculars or a noninverting telescope.

A-1b. You should be able to see more detail near quarter phase, as the Sun strikes the mountains on the Moon obliquely and casts readily visible shadows.

Answers to Selected Inquiries

A-1d. The quality of your drawing will depend on the size telescope used. With a telescope of aperture greater than a couple of inches, your drawing should be better than Galileo's.

A-2a. The center is brighter than the limb.

A-2b. Increase, since one sees deeper into the Sun in the center.

A-3a. The *shape* of the hole should have no effect.

A-3c. The largest hole casts the brightest image.

Suggestions for Further Reading
on Observational Astronomy

General Information on the Sky

"Graphic Timetable of the Heavens" is a graphical summation of the year's celestial events on one chart. Published in the January edition of *Sky and Telescope* magazine and also available from Graphic Ephemeris, P.O. Box 10561, Towson, MD 21204.

The Observer's Handbook, by Roy Bishop, Acadia University, Wolfville, Nova Scotia, Canada. Published yearly by the Royal Canadian Astronomical Society, this book supplies extensive lists on planets, stars, double and variable stars, deep-sky objects, occultations, comets and meteor showers, and Sun and Moon rise and set.

"Sky Calendar," from Abrams Planetarium, Michigan State University, East Lansing, MI 48824. A monthly calendar with day-by-day information. $7.50 per year.

"Star Date" is an informative 1- or 2-minute radio presentation keyed to the night sky on the day of the broadcast. Check to see if a radio station in your area carries the program. From the University of Texas McDonald Observatory.

Popular Magazines

Astronomy, Kalmback Publishing Co., 21027 Crossroads Circle, P. O. Box 1612, Waukesha, WI, 53187.

Sky and Telescope, Sky Publishing, P. O. Box 9111, Belmont, MA 02178-9111.

Star Date Magazine, from the Astronomy Department, University of Texas, Austin, TX 78712. Current skywatching information and short articles on astronomy. $15 per year.

Selected Star Maps and Sky Atlases

Burnham's Celestial Handbook. Dover, 1978. A 2138-page compendium.

Celestial Objects for Common Telescopes, by T. Webb. Dover, 1962. 2 volumes.

Field Guide to the Stars and Planets, 3rd ed., by J. Pasachoff and D. Menzel. Houghton - Mifflin, 1992. A useful primer on observing that contains star maps and white-on-black sky views.

Norton's 2000.0 Star Atlas and Reference Handbook. Sky Publishing, P. O. Box 9111, Belmont, MA 02178-9111. In its 18th edition, this is probably the most universally used star atlas by amateur astronomers.

Suggestions for Further Reading on Observational Astronomy

Sky Atlas 2000.0, by W. Tirion. Sky Publishing. The newest atlas on the market, attractively done.

Other Selected References for Amateur Astronomers

The following are a few of the many additional sources available. A more complete listing is the catalog of Sky Publishing, P. O. Box 9111, Belmont, MA 02178-9111.

All About Telescopes, by S. Brown. One of a set of useful pamphlet-books from Edmund Scientific Company. Other titles include: *Sky Log, Sky Guide, Time in Astronomy, Telescope Optics, Telescopes You Can Build, How To Use Your Telescope*, and *Mounting Your Telescope*.

Astronomy With Binoculars, by James Muirden. Arco, 1984.

Observational Astronomy for Amateurs, 4th ed., by J. Sidgwick. Enslow, 1982.

Selecting Your First Telescope, by S. Harrington. A pamphlet available from the Astronomical Society of the Pacific, 1290 24th Avenue, San Francisco, CA 94122.

Star-Hopping for Backyard Astronomers, by Alan MacRobert, 1994. Sky Publishing, Cambridge, MA.

Webb Society Deep-Sky Observers Handbook, K.G. Jones, Ed. Enslow, 1982. Five volumes with enough information to keep most observers busy for a lifetime.

The Whitney Star Finder, by C. Whitney. Knopf, 1977.

Star Maps

Index to the Star Maps for the Northern Hemisphere

Choose the date and time; the appropriate map is then specified.

Date	8 p.m.	9 p.m.	10 p.m.	11 p.m.	12 p.m.	1 a.m.
Jan-05		1		2		3
Jan-20	1		2		3	
Feb-05		2		3		4
Feb-20	2		3		4	
Mar-05		3		4		5
Mar-20	3		4		5	
Apr-05		4		5		6
Apr-20	4		5		6	
May-05		5		6		7
May-20	5		6		7	
Jun-05		6		7		8
Jun-20	6		7		8	
Jul-05		7		8		9
Jul-20	7		8		9	
Aug-05		8		9		10
Aug-20	8		9		10	
Sep-05		9		10		11
Sep-20	9		10		11	
Oct-05		10		11		12
Oct-20	10		11		12	
Nov-05		11		12		1
Nov-20	11		12		1	
Dec-05		12		1		2
Dec-20	12		1		2	

Star Maps

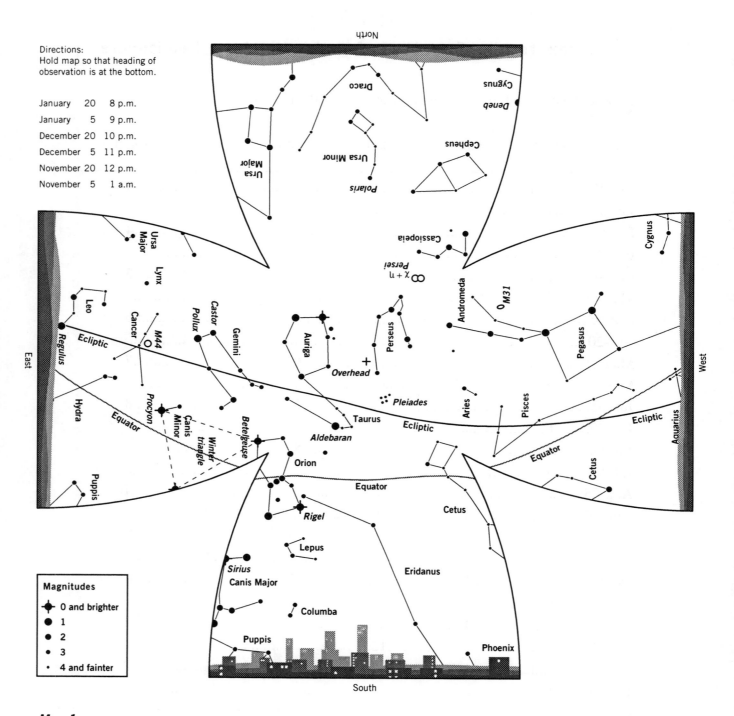

Directions:
Hold map so that heading of observation is at the bottom.

January	20	8 p.m.
January	5	9 p.m.
December	20	10 p.m.
December	5	11 p.m.
November	20	12 p.m.
November	5	1 a.m.

Magnitudes

✦ 0 and brighter
● 1
● 2
• 3
· 4 and fainter

Map 1

Star Maps

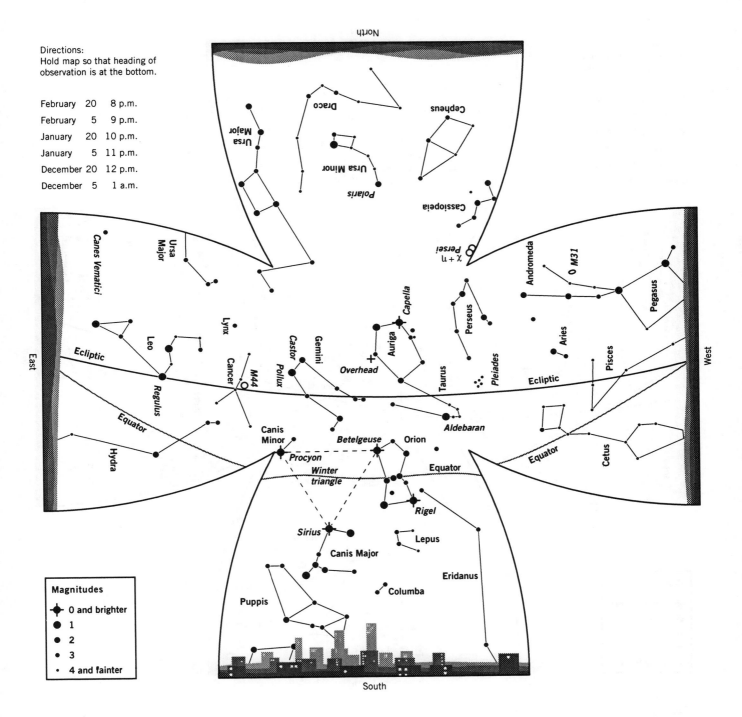

Directions:
Hold map so that heading of observation is at the bottom.

February	20	8 p.m.
February	5	9 p.m.
January	20	10 p.m.
January	5	11 p.m.
December	20	12 p.m.
December	5	1 a.m.

North

East

West

South

Draco

Ursa Major

Ursa Minor

Polaris

Cepheus

Cassiopeia

χ + h Persei

Canes Venatici

Ursa Major

Andromeda

0 M31

Pegasus

Lynx

Leo

Regulus

Ecliptic

Equator

Hydra

Cancer

M44

Castor

Pollux

Gemini

Overhead

Auriga

Capella

Taurus

Aldebaran

Perseus

Pleiades

Aries

Ecliptic

Pisces

Cetus

Equator

Canis Minor

Procyon

Winter triangle

Betelgeuse

Orion

Equator

Rigel

Lepus

Eridanus

Sirius

Canis Major

Columba

Puppis

Magnitudes

✦ 0 and brighter
● 1
● 2
· 3
· 4 and fainter

Map 2

Star Maps

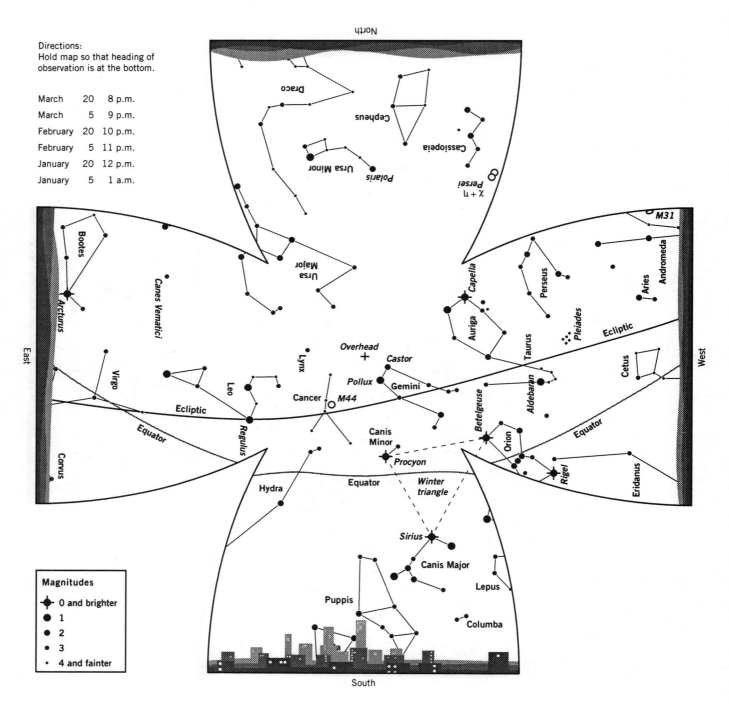

Directions:
Hold map so that heading of observation is at the bottom.

March	20	8 p.m.
March	5	9 p.m.
February	20	10 p.m.
February	5	11 p.m.
January	20	12 p.m.
January	5	1 a.m.

North

Draco

Cepheus

Cassiopeia

Ursa Minor

Polaris

Persei
χ + η

M31

Andromeda

Aries

Perseus

Capella

Auriga

Pleiades

Ecliptic

Taurus

Cetus

West

Aldebaran

Betelgeuse

Orion

Rigel

Eridanus

Equator

Bootes

Arcturus

Canes Venatici

Ursa Major

East

Virgo

Lynx

Leo

Cancer M44

Overhead
+

Castor

Pollux Gemini

Regulus

Ecliptic

Equator

Corvus

Hydra

Equator

Canis Minor

Procyon

Winter triangle

Sirius

Canis Major

Lepus

Puppis

Columba

Magnitudes
- ✦ 0 and brighter
- ● 1
- ● 2
- • 3
- · 4 and fainter

South

Map 3

Star Maps

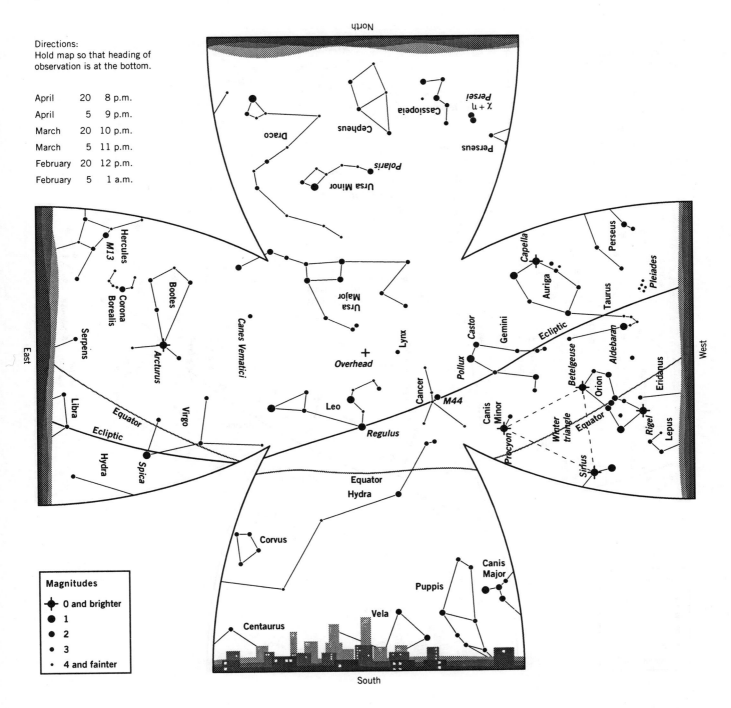

North

Draco

Cepheus

Cassiopeia

χ + η
Persei

Perseus

Polaris

Ursa Minor

Hercules
M13

Corona
Borealis

Bootes

Serpens

Arcturus

Ursa Major

Lynx

Canes Vematici

+
Overhead

Capella

Auriga

Perseus

Pleiades

Taurus

Gemini

Castor

Ecliptic

Aldebaran

Pollux

Betelgeuse

Orion

Eridanus

Cancer

M44

Leo

Regulus

Canis
Minor

Winter
triangle

Equator

Rigel

Lepus

Libra

Equator

Virgo

Ecliptic

Procyon

Sirius

Hydra

Spica

West

East

Equator
Hydra

Corvus

Canis
Major

Puppis

Vela

Centaurus

South

Magnitudes
- ✦ 0 and brighter
- ● 1
- ● 2
- • 3
- · 4 and fainter

Map 4

Star Maps

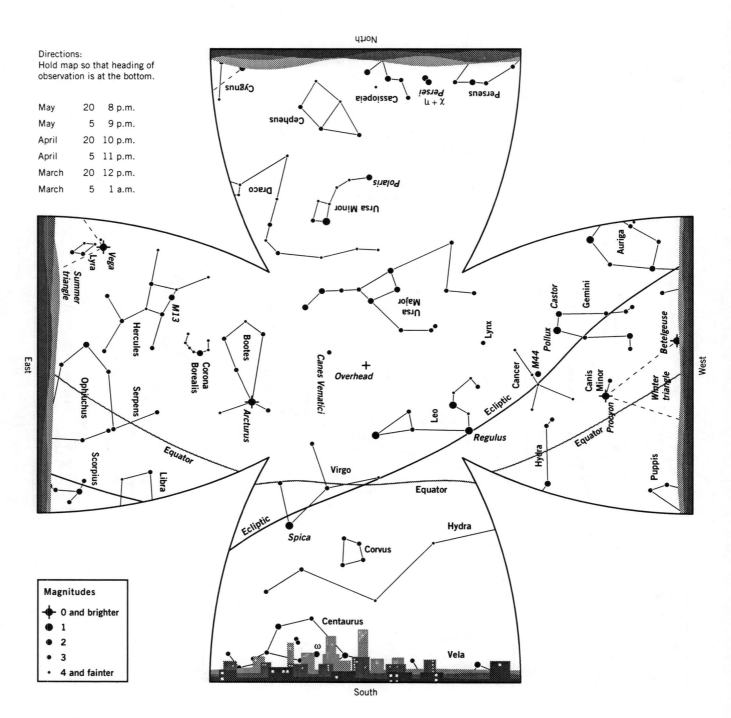

Directions:
Hold map so that heading of
observation is at the bottom.

May	20	8 p.m.
May	5	9 p.m.
April	20	10 p.m.
April	5	11 p.m.
March	20	12 p.m.
March	5	1 a.m.

North

Cygnus
Cassiopeia
χ + η Persei
Perseus
Cepheus
Draco
Polaris
Ursa Minor

East

Vega
Lyra
Summer triangle
M13
Hercules
Corona Borealis
Bootes
Ophiuchus
Serpens
Arcturus
Scorpius
Libra
Equator

Ursa Major
Canes Vematici
+ Overhead

Lynx
Castor
Gemini
Auriga
Pollux
M44
Cancer
Canis Minor
Betelgeuse
Leo
Ecliptic
Regulus
Hydra
Equator
Procyon
Winter triangle
Puppis

West

Virgo
Equator
Ecliptic
Spica
Corvus
Hydra
Centaurus
ω
Vela

South

Magnitudes
- ✦ 0 and brighter
- ● 1
- ● 2
- · 3
- · 4 and fainter

Map 5

Star Maps

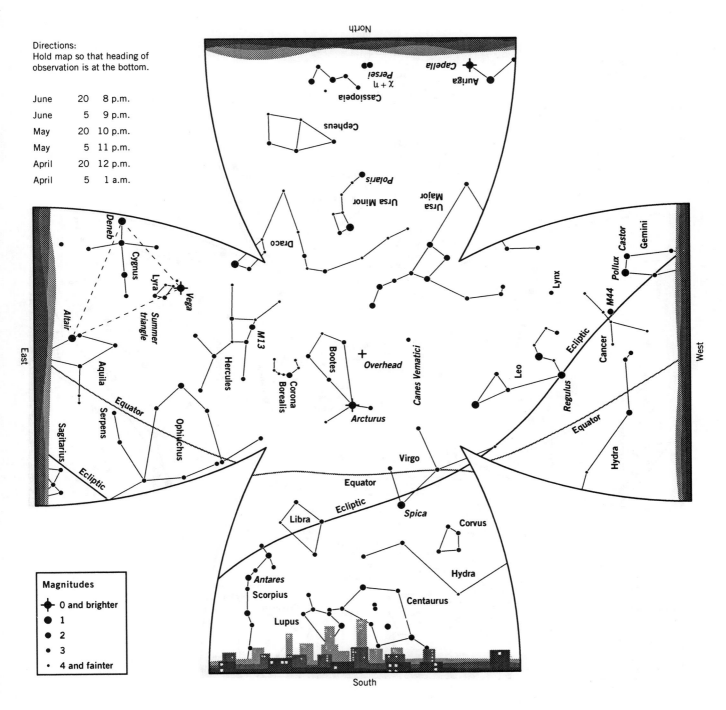

Directions:
Hold map so that heading of observation is at the bottom.

June	20	8 p.m.
June	5	9 p.m.
May	20	10 p.m.
May	5	11 p.m.
April	20	12 p.m.
April	5	1 a.m.

North

East

West

South

Magnitudes
- ✦ 0 and brighter
- ● 1
- ● 2
- ● 3
- · 4 and fainter

Map 6

Star Maps

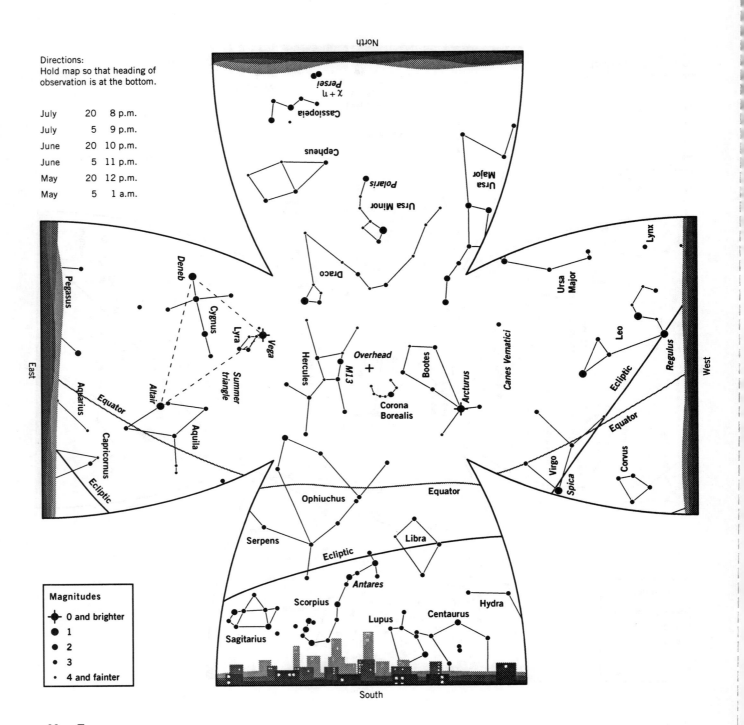

Directions:
Hold map so that heading of observation is at the bottom.

July	20	8 p.m.
July	5	9 p.m.
June	20	10 p.m.
June	5	11 p.m.
May	20	12 p.m.
May	5	1 a.m.

North

Cassiopeia

χ + η
Persei

Cepheus

Polaris

Ursa Minor

Ursa Major

Draco

Pegasus

Deneb

Cygnus

Lyra

Vega

Summer triangle

Altair

Aquila

Aquarius

Capricornus

Ecliptic

Equator

East

Hercules

M13

Overhead
+

Corona Borealis

Bootes

Arcturus

Canes Vematici

Ursa Major

Lynx

Leo

Regulus

Ecliptic

Equator

Virgo

Spica

Corvus

West

Ophiuchus

Equator

Serpens

Ecliptic

Libra

Antares

Scorpius

Lupus

Centaurus

Hydra

Sagitarius

South

Magnitudes

+ 0 and brighter
● 1
● 2
• 3
· 4 and fainter

Map 7

- 174 -

Star Maps

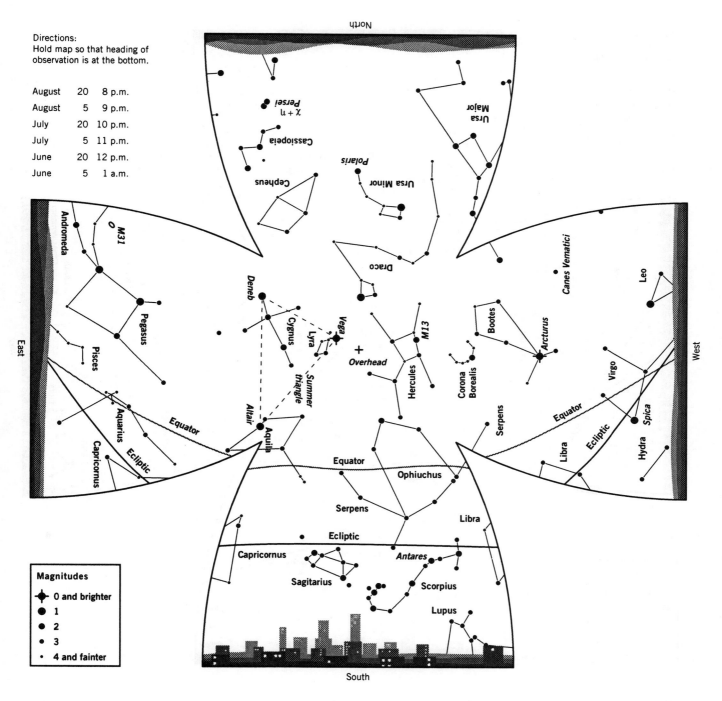

Directions:
Hold map so that heading of observation is at the bottom.

August	20	8 p.m.
August	5	9 p.m.
July	20	10 p.m.
July	5	11 p.m.
June	20	12 p.m.
June	5	1 a.m.

North

Persei χ + η
Cassiopeia
Cepheus
Polaris
Ursa Minor
Ursa Major
Draco

Andromeda
M31
O
Pegasus
Pisces
Aquarius
Capricornus
Equator
Ecliptic
Deneb
Cygnus
Lyra
Vega
Summer triangle
Altair
Aquila
Overhead +
Hercules
M13
Bootes
Arcturus
Corona Borealis
Serpens
Canes Venatici
Leo
Virgo
Libra
Equator
Ecliptic
Hydra
Spica
East
West

Equator
Ophiuchus
Serpens
Ecliptic
Libra
Capricornus
Antares
Sagittarius
Scorpius
Lupus

Magnitudes
- ✦ 0 and brighter
- ● 1
- ● 2
- • 3
- · 4 and fainter

South

Map 8

Star Maps

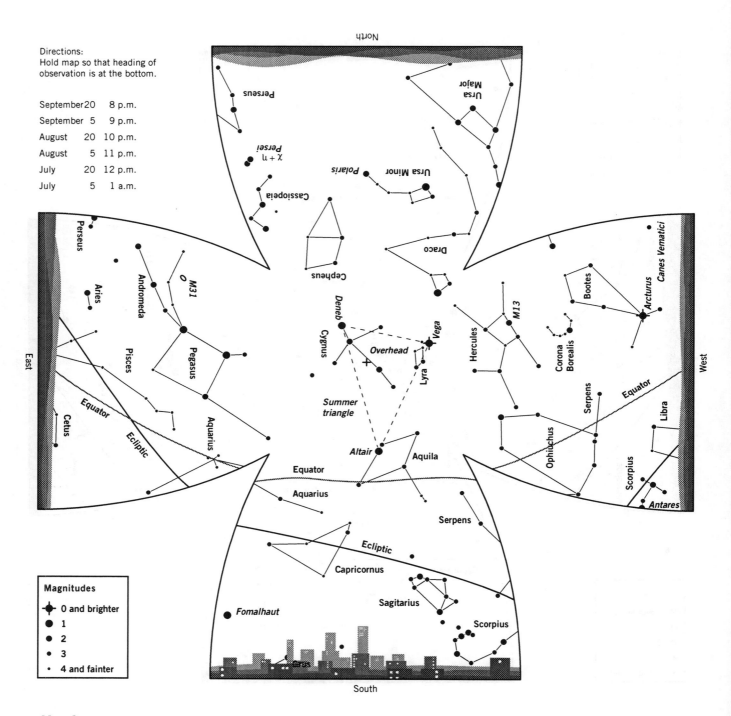

Directions:
Hold map so that heading of observation is at the bottom.

September	20	8 p.m.
September	5	9 p.m.
August	20	10 p.m.
August	5	11 p.m.
July	20	12 p.m.
July	5	1 a.m.

North

Perseus

Ursa Major

χ + η Persei

Polaris
Ursa Minor

Cassiopeia

Draco

Cepheus

Perseus

Aries

Andromeda

M31

Pisces

Pegasus

Cetus

Equator

Ecliptic

Aquarius

East

Deneb

Cygnus

Overhead

Summer triangle

Altair

Aquila

Equator

Aquarius

Vega

Lyra

Hercules

M13

Corona Borealis

Serpens

Ophiuchus

Bootes

Arcturus

Canes Venatici

Equator

Serpens

Libra

Scorpius

Antares

West

Equator

Serpens

Ecliptic

Capricornus

Sagittarius

Scorpius

Fomalhaut

Grus

South

Magnitudes

- ✳ 0 and brighter
- ● 1
- ● 2
- • 3
- · 4 and fainter

Map 9

Star Maps

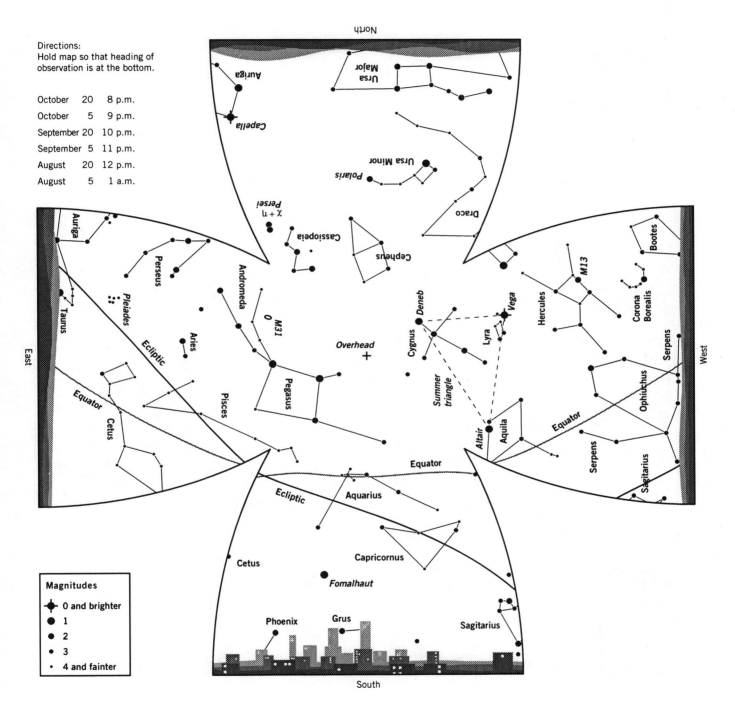

Directions:
Hold map so that heading of observation is at the bottom.

October	20	8 p.m.
October	5	9 p.m.
September	20	10 p.m.
September	5	11 p.m.
August	20	12 p.m.
August	5	1 a.m.

North

Auriga
Capella
χ + η Persei
Cassiopeia
Cepheus
Ursa Major
Ursa Minor
Polaris
Draco

East

Auriga
Perseus
Pleiades
Taurus
Aries
Andromeda
M31
Pisces
Pegasus
Ecliptic
Equator
Cetus

Overhead +

Cygnus Deneb
Vega
Lyra
Summer triangle
Altair
Aquila
Hercules
M13
Bootes
Corona Borealis
Serpens
Ophiuchus
Serpens
Sagittarius
Equator

West

Equator
Ecliptic
Aquarius
Capricornus
Cetus
Fomalhaut
Phoenix
Grus
Sagittarius

South

Magnitudes
✦ 0 and brighter
● 1
● 2
• 3
· 4 and fainter

Map 10

- 177 -

Star Maps

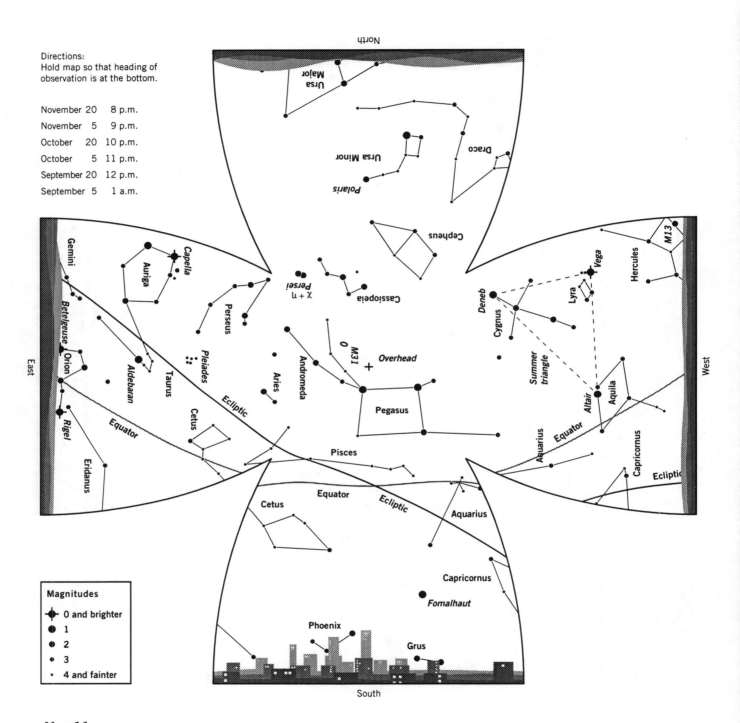

Directions:
Hold map so that heading of observation is at the bottom.

November 20	8 p.m.
November 5	9 p.m.
October 20	10 p.m.
October 5	11 p.m.
September 20	12 p.m.
September 5	1 a.m.

Magnitudes
- 0 and brighter
- 1
- 2
- 3
- 4 and fainter

Map 11

Star Maps

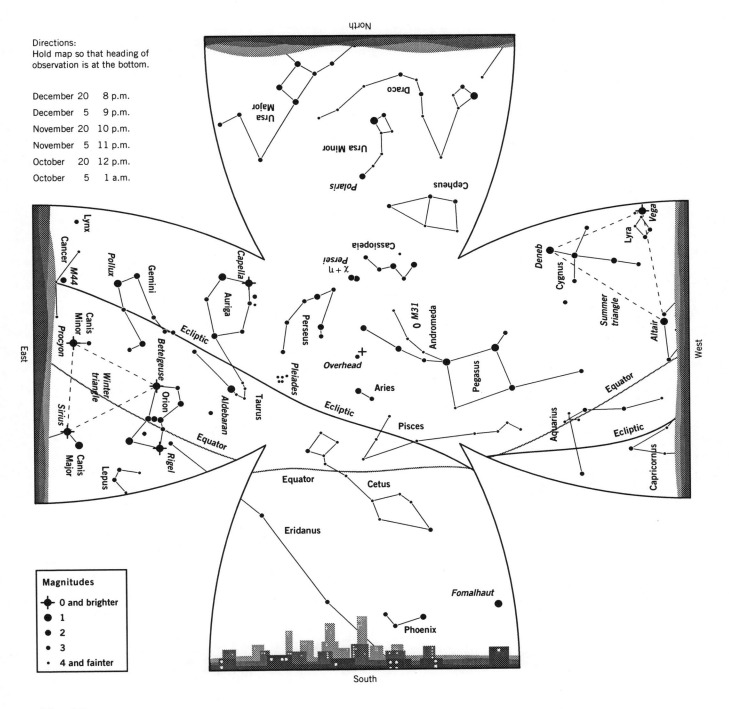

Directions:
Hold map so that heading of observation is at the bottom.

December	20	8 p.m.
December	5	9 p.m.
November	20	10 p.m.
November	5	11 p.m.
October	20	12 p.m.
October	5	1 a.m.

North

Ursa Major

Draco

Ursa Minor

Polaris

Cepheus

Cassiopeia

χ + η
Persei

Vega

Lyra

Deneb

Cygnus

Summer triangle

Altair

West

Lynx

Cancer

M44

Pollux

Gemini

Capella

Auriga

Canis Minor

Procyon

Betelgeuse

Ecliptic

Perseus

Pleiades

Overhead

Aries

Pisces

0 M31

Andromeda

Pegasus

Equator

Ecliptic

Aquarius

Capricornus

East

Winter triangle

Orion

Taurus

Aldebaran

Ecliptic

Sirius

Canis Major

Lepus

Rigel

Equator

Equator

Cetus

Eridanus

Fomalhaut

Phoenix

Magnitudes
- ✦ 0 and brighter
- ● 1
- ● 2
- • 3
- · 4 and fainter

South

Map 12

Photo Credits

3-D separations by Ray Zone, Los Angeles, CA. Venus photos provided by *Sky and Telescope Magazine* through the courtesy of NASA and Jet Propulsion Laboratory.

Figure 8-2-1: Courtesy Dr. Leon Kosofsky, National Space Science Data Center. Figure 8-2-2: Courtesy Dr. Leon Kosofsky, National Space Science Data Center. Figure 8-2-3: Courtesy Dr. Leon Kosofsky, National Space Science Data Center. Figure 10-1-1a: Courtesy Lowell Observatory. Figure 10-1-1b: Courtesy Lowell Observatory. Figure 19-1-1A: Courtesy Owen Gingerich, Center for Astrophysics, Cambridge, MA. Figure 19-1-1b: Courtesy Owen Gingerich, Center for Astrophysics, Cambridge, MA. Figure 19-1-2: Courtesy Lick Observatory. Figure 21-1-1: Courtesy Palomar Observatory. Figure 21-2-1a: Courtesy Palomar Observatory. Figure 21-2-1b: Courtesy Palomar Observatory. Figure A-1-1: Courtesy New York Public Library Picture Collection. Figure A-1-2: Courtesy New York Public Library Picture Collection.